64
GEEKS

Falkirk Community Trust	
30124 03102922 8	
Askews & Holts	
609.22	£12.99
MK	

An Hachette UK Company
www.hachette.co.uk

First published in Great Britain in 2018 by Ilex, a division of
Octopus Publishing Group Ltd
Carmelite House
50 Victoria Embankment
London EC4Y 0DZ
www.octopusbooks.co.uk
www.octopusbooksusa.com

Distributed in the US by Hachette Book Group
1290 Avenue of the Americas, 4th and 5th Floors, New York, NY 10104

Distributed in Canada by Canadian Manda Group
664 Annette Street, Toronto, Ontario, Canada M6S 2C8

Publisher, Photo and Tech: Adam Juniper
Editorial Director: Helen Rochester
Managing Editor: Frank Gallaugher
Senior Editor: Rachel Silverlight
Publishing Assistant: Stephanie Hetherington
Art Director: Julie Weir
Designer: e-Digital Design
Cover Design: Eoghan O' Brien
Senior Production Manager: Peter Hunt

ISBN 978-1-78157-572-7

A CIP catalogue record for this book is available
from the British Library.

Printed and bound in China

10 9 8 7 6 5 4 3 2 1

From Aristotle To Zuckerberg

CANCELLED

MEET THE BRAINS WHO CHANGED THE WORLD

Chas Newkey-Burden

ilex

CONTENTS

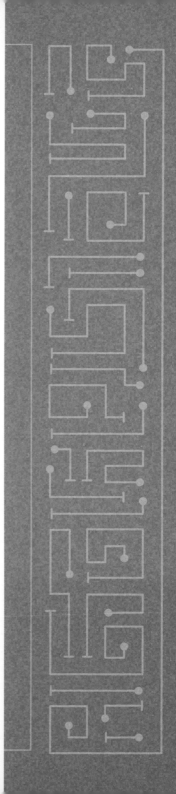

CHAPTER

1

CLASSICAL THINKERS & RENAISSANCE BRAINS

Pythagoras

Plato

Aristotle

Euclid

Archimedes

Leonardo da Vinci

Nicolaus Copernicus

Galileo Galilei

Blaise Pascal

Isaac Newton

Pythagoras

(570–495 BCE)

His teachings paved the way for the philosophies of Plato and Aristotle; he also had a huge impact on mathematics and is considered the first pure mathematician.

Mystery cloaks much of the life of this Ionian Greek philosopher. It is thought that Pythagoras was born on the Greek island of Samos around 570 BCE. Eager to escape the harsh rule of the tyrant Polycrates, he spent much of his youth traveling around various countries, including Egypt and Persia.

In 530 BCE, his itchy feet took him to Croton, in southern Italy, where he founded a school. There, his devoted pupils were lived a structured, ascetic lifestyle, based on tireless study and exercise. These followers became known as the Pythagoreans.

Although today best known for his mathematics, many of his teachings were of a spiritual nature. For example, he taught that perfection of the physical form in this life would allow immortality in the next. Transmigration of the soul, particularly reincarnation after death, was a radical concept in these times.

His most famous idea was the theory that the square of the hypotenuse (the side opposite the right angle) is equal to the sum of the squares of the other two sides. This is known as Pythagoras's theorem. It is one of the basic building block of maths, as it allows you to calculate the distance between two points. One of the finest achievements of Greek geometry, it goes beyond simple triangles into shapes of far greater complexity and more advanced mathematics and it is still used by everyone from video-game designers to builders and solar physicists, as well as every GPS system.

A big fan of numbers, Pythagoras argued for mathematics being the root of all things, developing the theory of the functional significance of numbers in the physical world and in music. He also thought numbers had personalities —he regarded each one as either masculine or feminine, perfect or incomplete, beautiful or ugly.

Around 500 BCE there was an uprising against the Pythagoreans and a rival group launched an attack. Pythagoras fled and it is believed he was killed or died shortly afterward.

BEST KNOWN TODAY FOR:
His eponymous theorem, which allows you to calculate the length of the third side of a right-angled triangle when only the lengths of the other two sides are known.

Plato

(c.427–c.347 BCE)

Although born an Athenian nobleman, after receiving a top-notch education under tutors as wise as Socrates and Cratylus, Plato rejected social privilege to devote his life to philosophy. It proved a judicious choice: his dazzling writings and piercing thought laid the foundations of Western thought as we know it today.

Given his antiquity, there is dispute over many details of Plato's life, but some details are widely agreed upon. It is known that—somewhat surprisingly—he fought in the Peloponnesian War between Athens and Sparta from 409 to 404 BCE. For a while Plato considered a career in politics but changed his mind after his mentor, Socrates, was executed in 399 BCE. The loss to politics was philosophy's gain. In an epochal and symbolic journey, he left Athens and spent several years traveling around the Mediterranean Sea, studying under different teachers, and writing.

When he returned, he formed the Academy, which became the most famous philosophical school. Among its famous alumni is Aristotle.

In some of the earliest surviving treatments of such issues, Plato wrote about a vast range of philosophical topics, including love, justice, beauty, politics, ethics, metaphysics, cosmology, and law. He was the innovator of the written dialogue and dialectic forms of philosophy.

His most famous work, *The Republic*, was a Socratic dialogue written around 380 BCE. In its pages he defined his concept of justice, as well as the features of a just city-state and characteristics of the just man. But he also posed challenging questions, asking, for instance, whether it is always better to be just than unjust.

The Republic included Plato's myth of the cave. This thought experiment proposed that humans are like men sitting in a cave seeing only shadows on the wall. He called for us all to step out of the cave and walk in the light—in other words to recognize and reject the dim distorted illusions of reality as we think we know it, and to seek instead full, technicolor reality.

The English philosopher Alfred North Whitehead summed up Plato's stature succinctly. He noted that "the safest general characterization of the European philosophical tradition is that it consists of a series of footnotes to Plato."

As for Plato himself, he once said that "wise men talk because they have something to say; fools, because they have to say something." On which note we will draw a line under this section.

BEST KNOWN TODAY FOR:
Being the author of philosophical works of matchless influence, the proponent of the dialectical method of inquiry, and launching the first university in the world.

Aristotle

(384–322 BCE)

His life became an endless quest for what scientists and philosophers everywhere continue to seek: the very principles that govern humanity and its universe. His writing still inspires and informs us, more than 2,000 years later.

Aristotle, whose magnificent mind would open the door to more than seven centuries of philosophy, was born in the city of Stagira, Chalkidiki, on the northern edge of Classical Greece. He joined Plato's Academy in Athens during his late teens and stayed there for two decades. He went on to form his own school, the Lyceum, in Athens, where he spent most of the rest of his life studying, teaching, and writing. Having himself been taught by Plato, he in turn tutored Alexander the Great—an impressive intellectual chain if ever there was one.

Aristotle's writings covered a host of topics including physics, poetry, theatre, music, logic, rhetoric, linguistics, politics, ethics, and zoology. He has made momentous and enduring contributions to nearly every aspect of human knowledge. A standout plank of his philosophy was his systematic concept of logic. He wanted to establish a universal process of reasoning that would, ultimately, allow man to learn everything. Key to this was his proposal that "when certain things are laid down, something else follows out of necessity in virtue of their being so." Philosophers now know this method of deduction as a "syllogism"—and his work contains the earliest known formal study of logic.

You can take countless examples of everyday wisdom and trace them back to Aristotle. For instance, the modern maxim that "those that can, do, those that can't, teach" is a, somewhat harsh, adaptation of his observation that "those that know, do—those that understand, teach."

Overall, his work amounts to the first comprehensive system of Western philosophy. In 322 BCE, he died of a digestive complaint but he lives on through his work. Every subsequent scientist and philosopher is in his debt, and any debate in the fields of logic, aesthetics, political theory, and ethics must pay tribute to him.

BEST KNOWN TODAY FOR:
Being the father of Western philosophy and remaining one of history's greatest thinkers in politics, psychology, and ethics.

Euclid
(c.325–c.270 BCE)

All authors like a good review, you know. Euclid
got a very favorable one when a reader observed
that "If his book failed to kindle your youthful
enthusiasm, then you were not born to be a scientific
thinker." The critic's name? Albert Einstein.

Euclid lived in Alexandria, Egypt around 300 BCE, during the reign of Ptolemy I, the Macedonian Greek general. Little is known about his life, but much is known about his work. To put it simply: if you have studied mathematics at school you can pretty much count yourself one of his students.

Indeed, Euclid's work, *Elements*, was the main textbook for teaching math (especially geometry) from its publication in 300 BCE until the early twentieth century. It has been described as the most influential textbook ever written. *Elements* represents the culmination of Greece's mathematical revolution. It compiled and elegantly explained all known mathematics of the era, including the work of Hippocrates, Pythagoras, Theudius, Theaetetus and Eudoxus. Readers were taught formulas for calculating the volumes of solids such as cones and pyramids. They were handed proofs about geometric series, perfect numbers, and primes, and useful algorithms.

Euclid also dealt with the basic properties of triangles, parallels, parallelograms, rectangles, squares, and circles. Plane geometry, geometric progression, and irrational numbers are covered lucidly. His treatise also included theorems on the properties of numbers and integers, which paved the way for number theory. In total, it consisted of 465 theorems and proofs. His system of rigorous mathematical proofs remains the basis of mathematics 23 centuries later.

Ptolemy, who personally sponsored Euclid, once asked him if there was a fast-track way to study geometry without poring over *Elements*. The maths genius is reported to have replied that there was "no royal road to geometry."

Euclid's clarity and rigor became the yardstick for subsequent mathematicians. Some have claimed that, next to the Bible, *Elements* may be the most translated, published, and studied, of all the books produced in the Western world.

The European Space Agency's Euclid spacecraft was named in his honor. Scheduled for launch sometime during 2020, the craft will observe billions of faint galaxies on a mission to establish why the universe is expanding at an accelerating pace.

"Euclid" is an anglicized version of the Greek name which means "renowned, glorious." Math geeks throughout the ages would surely feel this is an apt moniker.

BEST KNOWN TODAY FOR:
Forming many of the key concepts of geometry and writing one of history's most pored-over mathematical text books.

Archimedes
(287–212 BCE)

Archimedes' life story is one of plentiful legend. The most celebrated is his rumored cry of "Eureka"—meaning, "I have found it"—after he discovered the law of hydrostatics. Commonly known as "Archimedes' principle," this law states that a body immersed in fluid loses weight equal to the weight of the amount of fluid it displaces.

According to the familiar folklore, the Greek scientist made his discovery when stepping into his bath, and then ran naked through the streets in celebration. Many historians believe the latter part of the story is an invention, albeit a harmless and rather pleasant one.

Archimedes was born in 287 BCE in Syracuse, on the eastern coast of Sicily. Although he moved to Alexandria in Egypt for his education, he soon returned to Syracuse, where he spent most of the rest of his life. He was well connected there—on friendly terms even with the king, Hieron II.

So began his remarkable work. In mathematics, Archimedes used infinitesimals in a similar way to modern integral calculus by using the method of exhaustion to prove a treasure trove of geometrical theorems, including the area of a circle, and the surface area and volume of a sphere.

His achievements include coming up with the principle of the lever and his invention of the compound pulley and the hydraulic screw—also known as the Archimedes screw—for raising water from a lower to higher level.

When the Romans attacked Sicily in 214 BCE, Archimedes invented weapons for the defense of Syracuse. Among them was a catapult, but the anecdote that he beamed sunlight from a series of mirrors to burn the Roman ships besieging Syracuse is probably another Archimedes story worth taking with a pinch of salt.

Even the circumstances of his death in 212 BCE are disputed. It is known that he perished during the Second Punic War, when Roman forces under General Marcellus captured the city of Syracuse after a two-year-long siege. Some accounts have it that he was killed when he offended a Roman soldier by saying he was too busy with a maths problem to meet the general. Other historians say he was killed as he attempted to surrender to a soldier.

His tomb carried a sculpture of a sphere and a cylinder, representing his favorite mathematical discovery. Even in his death, he continues to educate.

BEST KNOWN TODAY FOR:
Discovering the law of hydrostatics and his celebratory cry of "Eureka," which has become the go-to catchphrase for innovators.

ARCHIMEDES

Leonardo da Vinci

(1452–1519)

Would you like to be one of history's most revered artists, sculptors, engineers, scientists, or inventors? That wouldn't be bad, would it? Well, Leonardo da Vinci managed to make a colossal contribution in all of these fields.

He was born out of wedlock in Vinci, Italy. Because he was illegitimate, he was denied a proper education and many career opportunities. Somehow, this seemed to give him extra curiosity, energy, and ambition. By the time he was 15, he was the apprentice of the painter Andrea del Verrochio in Florence. The teenager's art stunned all who saw it, including his mentor.

In 1482, his life took a turn to a more scientific path. He left the cultural bosom of Florence and moved to Milan. There, he began his innovative work, which led to him designing many things hundreds of years ahead of their times, including his flying machine, or ornithopter. His notes mentioned bats, kites, and birds as sources of inspiration. His sketches from the late fifteenth century would also go on to inspire the helicopter.

Elsewhere, he wrote that if a man had "a tent made of linen of which the apertures have all been stopped up," he would "be able to throw himself down from any great height without suffering any injury." He had just conceived the parachute, which would come in handy for the flying machines and thrill-seekers of the future. He also conceived presciently the bicycle, the tank, solar power, and scuba diving gear.

In 1499, he left Milan and traveled to cities like Venice and Rome where he concentrated more on his art, including his most famous work, the *Mona Lisa*. His drawing of the *Vitruvian Man* has become culturally iconic. Working in hospitals and medical schools, he honed his understanding of anatomy—another field in which he was far ahead of his time, and one which contributed to his art—through autopsies and dissections.

Whole books could be, and have been, written about this man's astonishing life. He died in 1519 at the age of 67. Ultimately, he had been neither scientist nor artist, but artist-engineer. His deepest scientific thoughts were expressed through art, and his harmonizing of these two fields is yet another of his many legacies.

Walking past a cave as a child, da Vinci had felt simultaneously terrified that a huge monster might lurk within and gripped with curiosity to discover what was inside. Rather than spending his life cowering outside that metaphorical cave, he had always bravely ventured within.

BEST KNOWN TODAY FOR:
His famous works of art, including *The Last Supper* and the *Mona Lisa*, and finding time for a string of inventions.

Nicolaus Copernicus
(1473–1543)

His heliocentric hypothesis shook the ideas of
Aristotle and Ptolemy, whose celestial model had
put Earth at the center of the universe, with the
sun, stars and other planets moving around it.
With his new model of the universe, Nicolaus
Copernicus kick-started the scientific revolution.

Copernicus was 10 when his merchant father died, but he enjoyed the encouragement of several mentors. His uncle, who was a priest, became his guardian and ensured that he received a good education. Copernicus studied at the Krakow Academy and then moved to Italy to study law at the University of Bologna. There, he met a mathematics professor, Domenico Maria de Novara, who fostered Copernicus's interests in geography and astronomy.

He was a bright student. As well as mathematics, he learned painting, astronomy, medicine, and canon law. It is believed that he spoke Latin, German, Polish, Greek, and Italian. It was during his university years that he dropped his birth name, Mikolaj Kopernik and replaced it with its Latin form: Nicolaus Copernicus.

When his uncle died in 1512, Copernicus moved to Frauenberg, in modern-day Germany, and took position as a canon in the church. The salary allowed him to delve deep into astronomy. Over several decades, he developed radical new theories, which eventually formed the basis of his mind-bending book, *De Revolutionibus Orbium Coelestium—On the Revolutions of the Celestial Spheres*.

The Roman Catholic Church was far from pleased. To suggest that the Earth was not the center of the universe was to undermine long-established religious dogma, making the world look worryingly small and unimportant in relation to the universe, and so the Church banned Copernicus's book after he died. However, smart astronomers approved—and many concluded it was the greatest work ever in its field. Today we use the phrase "Copernican revolution" when we speak of a breakthrough that completely upsets everything we think we know to be true.

Copernicus died of a stroke at the age of 70, in 1543. He had never married and had no children. Instead, he had devoted his life to service and science.

His first grave was lost for centuries but discovered again in 2005. His remains were reburied in Frombork Cathedral in 2010. His new tombstone is complete with a golden sun with six planets orbiting it.

BEST KNOWN TODAY FOR:
Establishing the heliocentric model of the universe, which shows that planets orbit the sun rather than the Earth.

Galileo Galilei
(1564–1642)

During his remarkable life, Galileo Galilei challenged some daunting figures, including Aristotle and the Catholic Church. Talk about courage, the plucky polymath must have had balls the size of the planets he observed.

Born in Tuscany in 1564, he read medicine, philosophy, and mathematics at the University of Pisa. He so loved the academic life that he stayed on as a professor, a role that allowed him to indulge his passion for experimentation. He studied a host of areas including speed and velocity, gravity and free fall, and the principle of relativity.

Galileo had an astonishingly bright mind. In 1609, he heard of the invention of the telescope in Holland. He had never seen an example of a telescope, but merely on the strength of the account he had heard, he built a superior version and used it to make a series of astronomical discoveries.

Peering through his telescope, he spied mountains and valleys on the surface of the moon. Then he noticed something that would put him on a collision course with ruling orthodoxy. According to the geocentric model that was prevalent at the time, the Earth was the center of the universe and all planets orbited. But Galileo discovered that Jupiter had four moons, which were orbiting Jupiter itself.

This suggested that—as Copernicus had argued before him—not everything orbited the Earth. Astronomers and Christians were mortified. Then, in 1613, he published theories about sunspots, undermining the Aristotelian doctrine that the sun was perfect.

In 1614, he was accused of heresy and two years later the Church banned him from teaching his theories. He was plunged into heretical hot water again in 1632, when he published his beliefs in a controversial book.

Summoned to appear before the Inquisition in Rome, he was condemned as foolish and absurd and sentenced to life imprisonment. The sentence was later reduced to permanent house arrest at his villa in southern Florence.

After suffering fever and heart palpitations, he died on January 8, 1642, aged 77. Although he had been condemned through much of his life, he is now considered the father of science and is immortalized in the opera section of Queen's epic hit *Bohemian Rhapsody*.

Galileo transformed the era of natural philosophy into one of modern science, and ushered in the scientific revolution. Not bad for a life's work.

BEST KNOWN TODAY FOR:
Countless astronomical contributions including discovering Jupiter's four largest moons and his championing of the heliocentric model of the cosmos.

Blaise Pascal

(1623–1662)

Blaise Pascal's father found one way of making money from numbers: he was a tax collector. The younger Pascal also had a way with figures, but his approach would leave quite a different legacy.

His was a childhood of challenges. His mother passed away when Pascal was a toddler. As a result, he became very close to his two sisters, Gilberte and Jacqueline. He also became a boy who seemed in a hurry to grab what could be gained from life.

At the age of 16, Pascal wrote a groundbreaking treatise on projective geometry. With his key contention he stated that if a hexagon is inscribed in a circle, or a conic, then the three intersection points of opposite sides lie on a line, which he called the Pascal line. This became known as Pascal's theorem. Many heavyweights of the time, including Descartes, found it hard to accept that a teenager could possibly have come up with such smart thinking.

Pascal was unfazed by the doubters. He opened a correspondence with Pierre de Fermat on probability theory. In their fascinating letters, they discussed many topics, including gambling. Pascal realized that there is a fixed likelihood of a particular outcome when it comes to the roll of the dice.

Perhaps his best-known legacy is that of Pascal's triangle. It is a neat tabular presentation of binomial coefficients, where each number is the sum of the two numbers directly above it. Although previous mathematicians had produced similar results in the past, Pascal gave the theory practical value by using it to help solve problems in probability theory. Indeed, he laid the very foundation for the modern theory of probabilities.

He was also one of the inventors of the mechanical calculator and he showed great perseverance in bringing it about. Over a three-year period he tried building 50 prototypes before eventually settling on a finished model. Over the following 10 years he then constructed 20 finessed machines. He called them Pascal's calculators, or Pascalines. Tax authorities became enthusiastic customers of the gadget.

His religious conversion began in 1646 and within a matter of years he had turned from writing mathematical treatises to penning theological works. He died in Paris on August 19, 1662. He was just 39 years of age.

BEST KNOWN TODAY FOR:
Inventing the digital calculator and roulette machines.

Isaac Newton

(1642–c.1726)

Isaac Newton had a formidable ego and fearsome temper. He believed that he was put on Earth to decrypt the word of God and uncover a single system of the universe.

Well, plenty of people think highly of themselves, but the genius physicist and mathematician, who was born on Christmas morning, has a life story that comes close to backing up his divine self-regard.

Discovering the laws of gravity and motion at a time when such things were a mystery, he would become a pioneer in setting a modern, rational worldview. Yet in his childhood years he had other preoccupations: "making pies on Sunday night ... punching my sister ... threatening my father and mother Smith to burn them and the house over them..."

At Cambridge University, he developed the philosophy behind his work—that making observations rather than studying books was the route to progress. He lived it to the full: one of his experiments involved sticking a blunt needle in his eye to see what happened.

This direct approach began to bear fruit: he revolutionized the telescope by using mirrors instead of lenses, creating a much smaller, and much more powerful, instrument. He also discovered that white light could be split into the colors of the spectrum, identifying the seven colors of the rainbow that English speakers around the world still remember with the acronym ROY G BIV (red, orange, yellow, green, blue, indigo, and violet).

Then he became preoccupied by alchemy, and revolutionary religious theories. These twin obsessions seemed to bend his mind to such an extent that, in 1687, it began to conjure groundbreaking concepts: of calculus, the three laws of motion, and a theory of gravity.

The latter theory came to him, he said, when, as he sat beneath an apple tree at his family home, a falling apple hit him on the head, prompting him to wonder why it fell straight down and not at an angle. The tale is arguably the most widely known legend in scientific history.

His life was not without conflict. When German philosopher Gottfried Leibniz announced a "new" theory of calculus, Newton angrily produced his notes from 20 years previously to prove that he had got there first. He later suffered a nervous breakdown, during which he wished dead the philosopher John Locke and the MP Samuel Pepys.

Perhaps the leading icon of the scientific revolution of the seventeenth century, he was knighted in 1705 by Queen Anne of England. He had radically shifted the way we understand the universe.

BEST KNOWN TODAY FOR:
Discovering gravity, the colors of the spectrum, inventing calculus, and having an apple fall on his head.

CHAPTER

2

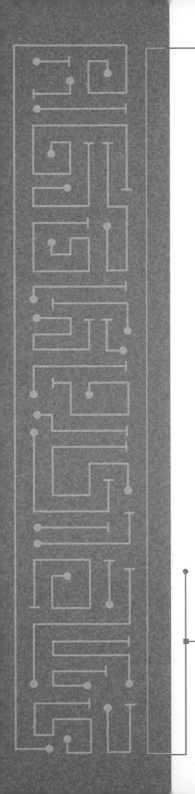

MACHINE-AGE MINDS

Benjamin Franklin

James Watt

Louis Daguerre

Charles Babbage

Charles Darwin

Ada Lovelace

Louis Pasteur

Alfred Nobel

Alexander Graham Bell

Thomas Edison

Sigmund Freud

Nikola Tesla

Konstantin Tsiolkovsky

The Wright Brothers

Benjamin Franklin
(1706–1790)

Benjamin Franklin was born in Boston on January 17, 1706. He left school early and helped out at his father's candle- and soap-making business, and at his brother's newspaper. After he and his brother fell out, Benjamin bought the *Pennsylvania Gazette* and became rich with his publishing success.

He used the money he made from his business ventures to move into scientific invention. One of his landmark inventions was the Franklin stove, which used a hollow baffle and inverted siphon to produce more heat and less smoke than existing units. Then came his lightning rod. After guessing that lightning was electricity, he came up with a plan to use a simple metal rod to conduct that electricity safely into the ground. With it, he hoped to save buildings from fires.

His famous kite experiment saw Franklin risk his life by flying a kite, fitted with a metal spike, directly under a thundercloud. Electricity ran down the kite's cord to a key tied near the end, creating a spark when he brought his hand close to it. The lightning rod was launched—it has since saved countless lives.

He also invented the bifocal spectacles. Frustrated from having to swap between his indoor and outdoor glasses, he simply cut them in half and joined them together in one frame. He could now see long distances through the top and read through the bottom. He innovated in countless other areas, too. For instance, he suggested that if people rose from their beds earlier, when it was lighter, then it would save on candles. He had just conceived the idea of changing the clocks twice a year.

All of this would already make for a rather marvelous resumé but Franklin went one further and became a founding father of the United States of America. In 1776, he helped to draft, and was then a signatory to, the Declaration of Independence.

Posted to France, he negotiated the Franco-American Alliance and then the Treaty of Paris, which ended the American War of Independence. After returning to America in 1785, he helped to draft the Constitution. He died in Philadelphia on April 17, 1790 at the age of 84.

BEST KNOWN TODAY FOR:
Being a founding father of the United States of America.

James Watt
(1736–1819)

The world's first working steam engine was patented in 1698, nearly 40 years before James Watt was even born, but thanks to his world-changing improvement of the engine, it is this Scotsman who is most associated with it.

Watt was born in Greenock, a poor seaport town in Scotland, on January 18, 1736. He was home schooled and showed manual dexterity, engineering skills, and a flair for mathematics early on in life. His childhood hobbies include taking apart and then reassembling his toys. Sometimes he built new playthings with the spare parts of ones he had deconstructed.

A colleague of Watt's father took one look at the boy's meticulous meddling and concluded, "He has fortune at his finger ends."

Watt began his working life as a maker of mathematical instruments but soon became obsessed with steam engines. Around 1764, he was given a model Newcomen engine to repair. He noticed how inefficient it was and designed a separate condensing chamber for the steam engine that prevented the enormous loss of steam it was suffering from.

His invention, which had come to him during a walk one Sunday afternoon, proved a defining moment of the Industrial Revolution thanks to its rapid incorporation into many industries. It was used to power machinery in factories. mines, and mills, calling time on industry's dependence on water power.

In 1775 Watt teamed up with Matthew Boulton, who owned a factory in Birmingham. Together, they began to manufacture steam engines. Boulton & Watt became the most important engineering firm in the country, meeting massive demand.

As Boulton put it: "I sell here, Sir, what all the world desires to have— power." Thank goodness one of them had a business mind. Watt had once said he would "rather face a loaded cannon than settle an account or make a bargain."

His influence was recognized when he was elected fellow of the Royal Society of London in 1785. He died on August 19, 1819 in Heathfield, England at the age of 83. He was buried alongside Boulton.

In 1882, the watt, a unit of measurement of electrical and mechanical power, was named in his honor. He has been described as "the most useful man who ever lived." Which is a very tidy eulogy.

BEST KNOWN TODAY FOR:
Pivotal enhancements in steam engine technology that powered the Industrial Revolution.

Louis Daguerre

(1787–1851)

The first photograph of a human being is believed to be an image of a man having his shoes shined on Paris's Boulevard du Temple. This is not the most dramatic of subjects, but when he took it, Louis Daguerre made history.

Born in 1787 in Cormeilles-en-Parisis, Val-d'Oise, France, Daguerre's education was disrupted by the French Revolution. Thanks to his flair at drawing, at 13 he was apprenticed to an architect, which allowed his creative, imaginative vision to be encouraged.

In the early 1820s, he had invented the diorama, a form of theater in which large transparent paintings were illuminated to simulate movement and other effects on a screen. The new genre, which opened in Paris in July 1822, saw audiences sit gobsmacked in special theaters.

However, he was then struck by a double-blow. First, a cholera outbreak in Paris meant that theater ticket sales sank. This was particularly damaging for diorama, which was expensive to produce. Then, his theater burned down. Overnight, his passion was denied him.

But fate was pushing him in a new direction. In 1829 he teamed up with Joseph Nicéphore Niépce, who four years earlier had produced the world's first permanent photograph. After 10 years of experimenting with chemicals and silver plates, Daguerre created and patented the daguerreotype process.

This was the first publicly available photographic technique, but it was an onerous process. The photographer first had to polish a sheet of silver-plated copper and treat it with fumes to make it light sensitive. They would then expose it in a camera for anywhere between a few seconds or many minutes, depending on the lighting conditions.

The image then would be made visible with mercury vapor and liquid chemical treatment, before being sealed behind glass in a protective enclosure. You wouldn't want to go through all that just to find out your subject had blinked at the crucial moment (or rather, fallen asleep—given how long each exposure took!).

He never truly cashed in on his invention, but the French government paid Daguerre an annual pension for publicly releasing it.

In the latter years of his life, Daguerre returned to his early passion: painting dioramas. He provided them to churches in and around the Paris suburb of Bry-sur-Marne, where he died on July 10, 1851, at age 63. The Eiffel Tower, built in the late 1880s, includes Daguerre's name inscribed on its base next to those of 71 other influential French scientists and inventors.

BEST KNOWN TODAY FOR:
The daguerrotype process: a precursor of modern photography.

Charles Babbage
(1791–1871)

Charles Babbage was often unwell as a child, so he was schooled mainly at home. This did not hold him back academically. In 1810, he went to Cambridge University where he pursued his passion for mathematics.

After graduating, he was hired by the Royal Institution to lecture on calculus. He was elected to the Royal Society and helped set up the Astronomical Society. Starting from 1828, he was Lucasian Professor of Mathematics at Cambridge and is considered pre-eminent among the many polymaths of his time.

In that same decade he created a six-wheeled machine that could perform mathematical calculations—he called it the "Difference Engine." He wanted to address the long-standing difficulty in producing error-free tables by teams of mathematicians and "human computers." Then he worked on another invention, which would be able to perform any arithmetical calculation using punched cards that would deliver the instructions.

Crucially, this machine developed concepts that would become fundamental elements of the computers that would emerge more than 100 years later. Among these were that the memory—known as the "Store"—was kept separate from the central processor—or the "Mill." It also included facilities for inputting and outputting data and instructions, and serial operation using a "fetch-execute cycle."

He called his device the Analytical Engine and by conceiving it he became a pioneer of modern computing. Ada Lovelace completed a program for the "Analytical Engine" but it was not finished in Babbage's lifetime. However, when the first general-purpose computers were built in the 1940s, his influence was felt across the planet.

He died on October 18, 1871, in London. Half of his brain is displayed at the Hunterian Museum in the Royal College of Surgeons in London. The other half of Babbage's brain is preserved in the Science Museum, London. What a brain it is.

BEST KNOWN TODAY FOR:
Essentially inventing the computer.

Charles Darwin
(1809-1882)

"Would it be too bold to imagine, that all warm-blooded animals have arisen from one living filament?" You would be forgiven for thinking those are the words of English naturalist Charles Darwin, but they were actually written by his grandfather Erasmus, who first posited the idea that one species could "transmute" into another.

Charles would take the idea into the mainstream. He studied medicine at Edinburgh University before moving to read Divinity at Cambridge. He was considering a career in the Church, which is ironic given his later battles.

Darwin's life, and the course of history itself, would change during a five-year voyage around the world, in which he studied variation in plants and animals. It was the best of times and the worst of times: he suffered seasickness "beyond what I ever guessed at" but also struck upon his theory of evolution via natural selection.

After observing species all over the globe, he came to believe that the species we know today had gradually evolved from common ancestors. In essence, his theory held that animals more suited to their environment survive longer and have more young.

But the idea that all species have evolved from simple life forms that first developed more than three billion years ago, and that humans shared a common ancestor with apes, put him on course for a collision with religious orthodoxy of Victorian times. He was threatening Creationism—the view that all of nature was born of God, and Darwin was aware of the storm he was about to cause.

After nervously keeping his theory secret for 20 years, he finally published *On the Origin of Species* in 1859. He said writing the book had been "like living hell," and releasing it to the public felt like "confessing a murder."

As he had predicted, the book sparked immediate and colossal controversy, but evidence supporting it would come in time. It was shown how the DNA of bacteria and viruses, including E. coli, could be damaged or altered during replication, sometimes to their benefit, such as allowing resistance to certain antibiotics.

His theory became the foundation of modern evolutionary studies and is now widely accepted and referred to simply as "Darwinism."

He died in 1882. Among his last words were to his wife, Emma. He told her, "I am not the least afraid of death. Remember what a good wife you have been to me. Tell all my children to remember how good they have been to me." He is buried at Westminster Abbey in London.

BEST KNOWN TODAY FOR:
His theories of evolution and natural selection that revolutionized human understanding of life on Earth.

Ada Lovelace

(1815–1852)

The first thing Lord Byron ever said to his daughter was: "Oh, what an implement of torture have I acquired in you!" Coming from a father this was harsh, but Ada Lovelace did indeed prove to be a challenging, complex individual—albeit one who would change the world.

The daughter of the poet and Lady Anne Isabella Milbanke Byron, Ada never knew her father, who separated from Lady Byron when their daughter was a matter of weeks old. Pushed by her mother toward mathematics and science in a bid to drive out any trace of her father's poetic waywardness, Ada hinted at her genius early. She showed a flair for numbers and language, and at the age of 12 she conceived the idea of an aeroplane.

In June 1833, at the age of 17, she met inventor/mathematician Charles Babbage, who became her mentor and dubbed her the "enchantress of numbers." Babbage had recently lost a daughter, who would have been about the same age as Ada.

He showed Ada two calculator devices he had invented: the Difference Engine and the Analytical Engine. While she was translating a Swiss article about the latter device, Ada added her own ideas about its potential flexibility and power. Crucially, she pinpointed the machine's potential to manipulate symbols rather than just numbers.

She also conjectured a method for the device to repeat a series of instructions—the looping process that computer programs still use to this day. Right there, she had prophesied modern computing a century in advance and become the world's first programmer. A historian describes how in her detailed notes Ada was "thumping the table" and highlighting hitherto unconceived-of capacity of the machine. Ada herself described her approach as "poetical science."

Had that poetical science been heeded faster, there could have been a computer revolution in the Victorian age. That it took 100 years for the significance of her notes to be recognized shows how visionary her intelligence was.

Ada was a thorny individual, who has been described as manipulative, aggressive, and bratty. She was a drug addict and keen gambler. But her exploits brought her into contact with scientific luminaries of the day, including Andrew Crosse, and the author Charles Dickens, who read to her on her deathbed.

Ada died of cancer in 1852 at the age of 36. She was buried alongside her father, whom she hard barely known during life, and who also passed away at the age of 36.

BEST KNOWN TODAY FOR:
The creator of the first computer program, Lovelace was the first person to recognize the full potential of what would become a computer.

Louis Pasteur

(1822–1895)

Characters like Superman, Captain America, and Wonder Woman save plenty of lives in the pages of comic books, but Louis Pasteur has saved millions of lives in the real world through his vaccines and insights into germs and food safety.

Born in 1822, Pasteur's childhood in the Jura mountains of eastern France seemed idyllic enough. As an artistic child, he loved to paint and draw. However, like many a budding creative, he was discouraged and told to focus on more academic pursuits.

This redirection would eventually serve him—and humanity—well. Taking a chemistry post at the University of Strasbourg, he showed that organic molecules with the same chemical composition could exist in either a left- or right-handed version, like a mirror image, taking on different behaviors according to their rotation. The 25-year-old Pasteur had revolutionized our understanding of DNA and paved the way for modern drug development.

Next up came his "germ theory," which swept away the "spontaneous generation" misunderstanding that had held sway since Aristotle's teachings. Showing that food went off because of contamination by microbes in the air, he posited that the same process could cause disease.

He then invented a way of heating drinks and food to keep them free from disease. The process, named pasteurization, was originally used to save French wine. Then he saved its silk industry from a devastating epidemic by teaching how infected silkworms should be destroyed.

On something of a roll, he injected bacteria into chicken and found they became immune to cholera. His discovery that weakened strains of a disease could prompt resistance to it was a game-changing contribution to the fight against infection.

He made a number of other breakthroughs, including the first human trial of a man-made rabies vaccine and the eradication of diphtheria. He also found time to relax every now and then. After all, he believed, "a bottle of wine contains more philosophy than all the books in the world."

BEST KNOWN TODAY FOR:
Creating the process of pasteurization and developing vaccinations against anthrax and rabies.

Alfred Nobel

(1833–1896)

He registered 355 patents but it was with one in particular that Alfred Nobel made the biggest bang. After his brother Emil was killed in a nitroglycerin explosion in 1864, Nobel set about trying to devise a way to more safely manufacture and deploy the volatile explosive.

Innovation was in his blood: Nobel's father was an engineer and inventor. During his late teens, Alfred traveled the world to study chemical engineering. He visited Sweden, Germany, France, and the United States.

Following his brother's death, Nobel got to work. It was a tough project: the trembling authorities banned experimentation with nitroglycerin within the city of Stockholm, so he moved his work to a barge, which was anchored on a lake. He eventually reached his goal by incorporating nitroglycerin into silica, which made it safer and easier to manipulate. He had just invented dynamite. He demonstrated the new explosive in 1867, at a quarry in Redhill, Surrey.

Nobel also developed a detonator that could be ignited by lighting a fuse, paving the way for the iconic dynamite stick of *Roadrunner* and other cartoons. It changed the real world, too. Although it can conjure a feeling of aggression, dynamite was quickly used for positive purposes: blasting tunnels, cutting canals, and building railways and roads. With the diamond-drilling crown and the pneumatic drill also coming into use at this time, such work became drastically simpler. As such, Nobel's invention became a nice little earner. He opened a network of factories across Europe to manufacture explosives. Then, in 1894, he bought an ironworks at Bofors in Sweden that soon became the hub of the Bofors arms factory.

His success lay in his knack to blend the piercing, innovating mentality of the inventor with the zip of the entrepreneur. A tireless man, he also built bridges and buildings in Stockholm.

But in 1888, when his brother Ludvig died, a French newspaper accidentally published an obituary of Alfred, describing him as the "merchant of death" due to his profits from the sales of arms.

Horrified that this might prove to be regarded as his eventual legacy, Nobel drew up his will to establish the now iconic Nobel Prizes. A massive slice of his personal fortune was used to institute annual prizes in Physics, Chemistry, Physiology or Medicine, Literature, and Peace. An Economics Prize was added later.

ALFRED NOBEL

BEST KNOWN TODAY FOR:
Inventing dynamite and devising the Nobel Prizes, the annual recognition of academic, cultural, and scientific progress.

Alexander Graham Bell

(1847–1922)

In a variety of ways, speech was a big thing in Alexander Graham Bell's family. His mother was almost deaf; his father taught deaf people to speak; his grandfather was an elocution buff. Yet these details hardly add up to the colossal impact that Bell himself would have on human conversation.

As a child, Bell would help his mother follow conversations by tapping a code out on her arm. He soon became obsessed with the idea of transmitting speech, and in the 1870s he made it happen. In 1871 he moved from his birthplace, Edinburgh, in Scotland, to America. There, he founded a technique to teach deaf-mute children, known as "visible speech." In Boston the following year, he established a school to train teachers of the deaf. He was appointed professor of vocal physiology at Boston University in 1873.

Two years later, he created a simple receiver that could turn electricity into sound—better known as a telephone. His reasoning had been simple: he had thought that rather than sending a code along an electrical wire, it would be better to send the actual sound of a human voice along a wire. He was granted a patent for the telephone in 1876, and, within a year, the first telephone exchange was built in Connecticut.

Bell made a lot of money from the invention because he owned a third of the shares in the Bell Telephone Company, created in 1877. This only fanned the flames of the controversy over whether he should be credited with the invention. Others, including an Italian-American called Antonio Meucci, also made claim to it. Today Meucci, who was handicapped by poverty, has been acknowledged for his contribution to the invention.

In 1880, Bell founded the Volta Laboratory in Washington, where he continued to experiment. He played around with ideas in a number of fields, including communication and medical research. He was awarded the French Volta Prize. In 1888, he was one of the founding members of the National Geographic Society. He served as its president from 1896 to 1904. Later, he bought up some land in Nova Scotia and built a summer home, where he could continue to experiment, including in aviation. He died there on August 2, 1922.

Ironically enough, he personally considered his big invention an unwelcome intrusion on his real vocation as a scientist. He refused to have a telephone in his study.

ALEXANDER GRAHAM BELL

BEST KNOWN TODAY FOR:
Inventing the telephone.

Thomas Edison

(1847–1931)

What did Thomas Edison ever invent? A better question would be, what didn't he invent? Most widely remembered as the man who made cheap electric light possible, the restless American was a serial inventor who patented a grand total of 1,093 devices.

Born in 1847 in Ohio, Michigan, at the age of 12 Edison printed and sold newspapers to train passengers. He began his career proper in telegraphy, which was then a fledgling sector. He was still in his early twenties yet he quickly made a name for himself with significant improvements to the technology of the telegraph.

His first two inventions were an electronic vote-recording machine and a stock ticker. Then his talent for innovation went into overdrive: he invented a new form of telephone microphone, the phonograph, and the light bulb. Some would have had their curiosity satisfied and their imaginative energy drained by this point, but not Edison.

In the 1890s, he made his presence felt in cinema, with the invention of the motion picture camera. His many inventions contributed hugely to mass communication and telecommunications, but he also made an impact in energy when he developed a system of electric-power generation and distribution to homes, businesses, and factories.

Edison was the creator of the world's first industrial research laboratory, where he encouraged an informal atmosphere for him and his colleagues, or "muckers" as he referred to them. "Hell, there ain't no rules in here, we're trying to accomplish something," he said at his purpose-built lab, which sat on a 34-acre site in Menlo Park, New Jersey.

He was as good as his word: he and his muckers enjoyed weekend drinking and sing-along sessions at the laboratory organ. However, he was also a demanding taskmaster who could "wither one with his biting sarcasm or ridicule one into extinction," remembered a colleague.

He held fascinating views, which he expressed memorably. "Gold is a relic of Julius Caesar, and interest is an invention of Satan," he once said. Asked about his views on religion, he said, "I do not believe in the God of the theologians; but that there is a Supreme Intelligence I do not doubt."

Edison died of complications related to diabetes on October 18, 1931, in his home in West Orange, New Jersey. A test tube at The Henry Ford museum near Detroit is believed to contain his last breath.

BEST KNOWN TODAY FOR:
Inventing the phonograph, the motion picture camera, the electric light bulb, and so much more...

Sigmund Freud
(1856–1939)

In 1933, the Nazis publicly burned books written by Sigmund Freud. If only Adolf Hitler had hopped onto Freud's analytical couch, the world might have been saved a lot of trouble. Few people were in a better position to analyze the impulses that lay behind the Führer's violent acts.

After studying medicine at the University of Vienna, Freud worked at the General Hospital where he treated hysteria by encouraging patients to recall painful and traumatic experiences under hypnosis. After a spell studying neurology in Paris, he returned to Vienna in 1885 and set up a private practice, specializing in nervous and brain disorders.

Deducing that people have an "unconscious" aspect to their psychology, where sexual and aggressive impulses rise up in conflict with more rational consciousness, he founded psychoanalysis. His method for treating mental and emotional disorders through a dialogue between patient and a psychoanalyst that uncovered these repressed instincts, known more casually as the "talking cure," was revolutionary for its time.

So was his analysis of dreams in terms of unconscious desires and experiences, an approach he developed in 1900, when he published *The Interpretation of Dreams*. Freud also devised the famous Oedipus complex, his theory of a child's latent sexuality, focused on the parent of the opposite sex.

Along with Carl Jung, he launched the International Psychoanalytic Association. In 1923, he published a paper entitled "The Ego and the Id." Over its mind-bending pages, he tore apart conventional theories about the structure of the mind, arguing that it was divided into the "id," the "ego," and the "superego"—respectively, and in the simplest terms, the instinctive, the rational, and the moralistic sides of consciousness.

Freud was not one to downplay the importance of his work. He pinpointed three revolutionary upheavals in thought: Nicolaus Copernicus and his heliocentric theory, which diminished Earth itself; Charles Darwin, who argued that man wasn't a Godly creation but had descended from apes; and his own theories which showed that the ego "is not even master in his own house."

He may not have been entirely wrong: though his theories can appear outdated to us today, they were hugely challenging and controversial in their day, and his work really did change the way we understand human psychology. And despite that dangerous-sounding level of self-regard (quite possibly fueled by his cocaine habit), Freud was equally frank as to the hard work, patience, and humility needed for human progress. He said: "From error to error one discovers the entire truth."

BEST KNOWN TODAY FOR:
Psychoanalysis. His influential theories including dream interpretation and the Oedipus complex.

Nikola Tesla

(1856–1943)

He was the father of the electrical age, he filed more than 700 patents and is now regarded as the patron saint of geeks—but Nikola Tesla died in anonymity and penury.

Tesla was influenced as a child by his mother, who invented several household appliances in her spare time. At the age of 28, he moved to America where, clutching a letter of recommendation, he presented himself to Thomas Edison, the famed inventor and the brain behind the DC-based electrical system.

Although Edison's DC (direct current) worked well with lighting, it could not transmit electricity for long distances. Tesla's AC (alternating current) system, however, could send electricity for thousands of miles using high voltage.

The future rivals worked together for several months before falling out. For a while, Tesla struggled, but in 1887 he secured funding to develop his AC electrical system.

This put him head-to-head with Edison and the two battled to sell their rival systems to the nation. It was a bitter feud: Edison waged a smear campaign against AC and arranged public electrocutions of an elephant to scare people off it. But Tesla prevailed. He was invited to demonstrate his system at the 1893 World's Columbian Exposition in Chicago and people liked what they saw. Within years, the first AC hydroelectric power plants were opened at Niagara Falls and Buffalo, New York.

Having won the "current wars," Tesla continued to innovate. He invented the "Tesla coil," which used polyphase alternating currents to create a transformer that could produce very high voltages and new frequencies, and which was also capable of sending and receiving radio waves. With its crackling sparks and sheets of electric flame it blew people's minds.

In his lab, he developed and used fluorescent bulbs some 40 years before they were officially invented. He was also a pacesetter in x-rays, radar, robotics, and radio remote control. It was said by associates that many of his innovations would arise from a momentary inspiration.

In 1901, he told the banker J. P. Morgan that he had a plan for a mobile messaging service, in which telegram messages would be funneled through a laboratory where they would be encoded and given a new frequency, which would then be sent to a handheld device. Sound familiar?

Morgan cut the funding, however, and the project was abandoned. Tesla suffered a nervous breakdown and increasingly began to withdraw from the world. He died in 1943 in the New Yorker Hotel. He was 86, reclusive, bankrupt, and, despite his enormous contribution to humankind, unacknowledged. He is now a cult hero.

BEST KNOWN TODAY FOR:
Mains electricity is all Tesla's AC. Elon Musk's car company is named after him.

Konstantin Tsiolkovsky
(1857–1935)

The fifth of 18 children born to a Polish
deportee to Siberia, at the age of 10
Konstantin Tsiolkovsky became virtually deaf
from scarlet fever and had to quit school.
When he was 13, his mother died.

Sometimes people never bounce back from childhood traumas and Tsiolkovsky had known more than his fair share of hardship. The brilliant and resilient Tsiolkovsky, however, went on to become the father of theoretical and applied cosmonautics, and developed insights that are still in use over a hundred years later.

After his early setback, he was homeschooled and later became a teacher. In his spare time, he loved to read science fiction, including Jules Verne's tales of space travel. He began to write his own galactic stories, which showed acute understanding of technology and science. His writing began to move from fiction to theoretical papers on liquid propellant rockets, gyroscopes, escape velocities, and the principle of action and reaction.

He built a centrifuge to test gravitational effects, using some unfortunate chickens as his test subjects. He published writing about the effects of zero gravity and then developed Russia's first wind tunnel.

A magazine article he wrote in 1903 laid the basis for spacecraft engineering. He established the mathematical relationship between the changing mass of a rocket as it burns fuel, the velocity of the exit gases, and the final speed of the rocket. That relationship became known as the Tsiolkovsky formula and laid the foundation of astronautics.

Underpinning his work was his fascination with the possibility of space travel that had begun in his teenage years, and a philosophy that humanity should be dominant in space, known as anthropocosmism. Explaining his approach, he said, "First, inevitably, the idea, the fantasy, the fairy tale. Then, scientific calculation. Ultimately, fulfillment crowns the dream."

He died at home on September 19, 1935 at the age of 78. A moon crater is named in his honor and so is the asteroid 1590 Tsiolkovsky.

"Earth is the cradle of humanity, but one cannot remain in the cradle forever," he once said. A fine metaphor for his revolutionary work and his restless, stratospheric life.

BEST KNOWN TODAY FOR:
Laying the groundwork for space travel long before it was a thing.

The Wright Brothers
(1867-1912 & 1871–1948)

The flight lasted 12 seconds and the craft traveled 120 feet at an altitude of just a couple of feet. Yet Orville Wright had just made history. He and his brother Wilbur had ushered in an era of frustrating queues, hidden charges, and appalling food. The age of aviation had dawned.

Their story begins in Iowa, a quarter of a century earlier, in 1878, when Milton Wright gave his two sons a toy helicopter powered by an elastic band. Wilbur, 12, and Orville, 8, were inspired. As Orville explained of their childhood home, "There was always much encouragement to children to pursue intellectual interests; to investigate whatever aroused curiosity."

The next chapter came in 1893, when the brothers opened a bicycle store. They analyzed what made a good bicycle—a frame that was sturdy yet light, with balance and wind resistance —and applied these things to their big dream: realizing air travel.

In 1899, Wilbur wrote to Washington's Smithsonian Institution, expressing that dream. "I am an enthusiast, but not a crank. I wish to avail myself of all that is already known and then, if possible, add my mite to help ... attain final success."

The brothers built three gliders and tested them at Kitty Hawk in North Carolina. When they were ready to attempt a manned flight, a series of obstacles got in their way, including poor weather, technical issues, and their bishop father's disapproval of them flying on Sundays. To up the stakes, they had a rival: Samuel P. Langley, secretary of the Smithsonian Institution, had constructed his own craft and was keen to beat the brothers to it.

On December 14, 1903, the Wrights tossed a coin to decide who would pilot their one-man plane, the Flyer. Wilbur won and was at the helm as it rolled down the launching rail ... then stalled and crashed.

But three days later, they succeeded. After repairing the Flyer, Orville was the pilot. At 10.35am, and despite a 27 mph wind, he climbed onto the Flyer and made the historic flight.

Humankind has dreamed of flying since ancient times. Images have been found scratched on stone pyramids. Greek mythology tells the tale of Icarus, son of the inventor Daedalus, whose artificial wings carried him too close to the sun. In Renaissance times, Leonardo da Vinci made drawings of flying machines. But it was Orville and Wilbur who made it happen. Not bad for the only members of the Wright family who did not attend college.

BEST KNOWN TODAY FOR:
Inventing, building, and flying the world's first successful airplane.

CHAPTER

3

WARTIME WHIZ KIDS

Marie Curie

Guglielmo Marconi

Albert Einstein

Fritz Pfleumer

Alexander Fleming

Emmy Noether

Percy Shaw

Jon von Neumann

J. Robert Oppenheimer

Grace Hopper

Alan Turing

Hedy Lamarr

Marie Curie
(1867–1934)

Perhaps the best-known female scientist of all time, Marie Curie was bitterly resisted by some male scientists and never earned proper income from her discoveries. She eventually paid the highest price for her pioneering work, but her legacy is extraordinary.

Born Marie Sklodowska in Warsaw on November 7, 1867, she was the youngest of five children of poor teachers. In 1891, she moved to Paris to study physics and mathematics. She was penniless but rich in zeal.

In the city of love she met a physics professor called Pierre, whom she married in 1895. As scientists, they formed an epochal collaboration: they discovered two elements, polonium and radium. Polonium was named after Marie's homeland, Poland; radium was named after the Latin word for ray. These discoveries laid the road for modern radiation therapy treatments of cancer and other illnesses. Their work also led to a greater understanding of how atoms are built up. The Curies were awarded the Nobel Prize for Physics in 1903.

Triumph turned to tragedy when, three years later, Pierre was knocked down by a horse-drawn carriage and killed. Marie vowed to devote her life to continuing the work that they had begun together.

She received a second Nobel Prize, for Chemistry, in 1911. But perhaps her finest, and certainly most dramatic, moment was yet to come. Marie was living in Paris when the First World War broke out. She realized that the electromagnetic radiation of X-rays could help doctors find bullets and shrapnel in the bodies of wounded soldiers.

She helped equip ambulances with mobile X-ray equipment, some of which she drove to the front lines herself. Historians say her mobile units helped save the lives of a million soldiers. She also offered to donate her gold Nobel Prize medals to the French government to aid the war effort, but her offer was rejected.

She became one of the first celebrity scientists, drawing the fascination of the news cameras and tabloid gossip; never more so than when she embarked on a passionate affair with a married scientist. The French press lapped up the tale.

On July 4, 1934, at the age of 66, she died from leukaemia, which was almost certainly caused by exposure to high-energy radiation. She was the first woman to win a Nobel Prize, the first person and only woman to have won twice, and the only person to win a Nobel Prize in two different sciences.

BEST KNOWN TODAY FOR:
Discovering radium and polonium, which became game-changers in the fight against cancer.

Guglielmo Marconi

(1874–1937)

Some people read page-turning action novels or raunchy romances on vacation, but Guglielmo Marconi preferred to spend his holidays poring through books about science.

Thank goodness for that because in 1894, while holidaying in the Alps, a book he was reading inspired Marconi to explore whether electromagnetic waves could send messages through the air in the same way as they could be sent along wires.

He chased that thought and went on to effectively invent the radio.

Following his Alps sojourn, he set up a transmitter that sent out radio waves. Nine meters away, he had placed a receiver connected to a bell. When he pressed a switch on the transmitter, it sent out electromagnetic waves that were picked by the receiver. The ringing of the bell confirmed his success.

His next success was sending messages in Morse code between a transmitter and receiver nearly one mile apart. He also developed what became Marconi's law: the relation between height of antennas and maximum signaling distance of radio transmissions.

He was granted the world's first wireless telegraphy patent and by 1901 he succeeded in transmitting the letter "s" in Morse code by radio telegraph across the Atlantic from Cornwall, England, to Newfoundland in Canada. He had confounded those who assumed that wireless messages would be compromised by the curvature of the planet.

Regular radio broadcasts began during the 1920s. By then, Marconi had narrowly avoided death when he almost boarded the Titanic for her maiden voyage. Those who survived the disaster did so because of his wireless, which was used to raise the alert.

His impact on the world continued long after his death in 1937. In 1962, the Goonhilly Earth station was opened in Cornwall, close to where he had sent his first transatlantic signal. It relayed radio from space satellites for live television broadcasts. In 2001, satellite phones allowed people to communicate from far-flung war zones.

He is the recipient of many honors. In 1909, he shared the Nobel Prize for Physics with Professor Karl Braun. The asteroid 1332 Marconia is named after him. So is a large crater on the far side of the moon.

BEST KNOWN TODAY FOR:
Inventing the radio and broadcasting the first transatlantic radio signal.

Albert Einstein

(1879–1955)

Clearly, the world has much to thank Albert Einstein for, but his parents also deserve a hat-tip for nurturing his independent, questioning nature. For instance, when he was four, they sent him out to explore the local area on his own.

Born in Ulm, Germany in 1879, Einstein grew up in Munich. At school, he rebelled against the rote style of study, leading his exasperated schoolmaster to declare, "He will never amount to anything." Not only did he fail his first entrance exam for a Zurich technical college, but once he got in he played truant and got his girlfriend pregnant.

Then he began to change the world. He published a paper showing that measurements of space and time were relative to motion. This theory, which became known as Einstein's Special Theory of Relativity, challenged basic concepts of physics.

In another paper he explained photoelectric effect, showing that light was not just made up of waves, it could also be thought of as individual particles, or photons.

His most significant thought experiment came when he pictured a man falling off a roof and realized he would not feel his own weight. This led him to the General Theory of Relativity. Based on the idea that space and time could not be thought of independently of each other, Einstein's theory proposed that gravity was the effect of matter bending this "spacetime."

Einstein had quite the way with words. He said the origins of all technical achievements were "divine curiosity and the play instinct." Coming at it from the opposite direction, he once wrote that "unthinking respect for authority is the greatest enemy of truth." In a more mischievous moment, he declared that two things are infinite: "the universe and human stupidity."

A lifelong pacifist, he said that peace cannot be achieved through violence and he refused to sign a manifesto defending Germany's empire-building at the start of the First World War. In 1939 he wrote to US President Roosevelt encouraging the development of atomic weapons, a move he later regretted. The FBI opened a 1,500-page file on him.

The ultimate personification of the eccentric genius, the crazy-haired Einstein is an enduring icon. He was awarded the 1921 Nobel Prize in Physics. He became a professor four times and was given honorary degrees from several universities including Oxford and Cambridge. He has appeared on the cover of *Time* magazine six times, and was its "Person of the Century" in 1999.

BEST KNOWN TODAY FOR:
His theory of relativity and the equation $E=MC^2$.

Fritz Pfleumer

(1881–1945)

Music mix tapes, gripping murder
confessions, boring college lectures
and so much more—the tape recorder
has captured a wide range of sounds
since it was invented in 1932.

The German-Austrian engineer responsible for the invention was a reclusive, mysterious man. Fritz Pfleumer made ends meet by selling innovative ideas as an industrial consultant. One such was a process for keeping metal stripes on cigarette papers. In 1920s Europe, gold-tipped cigarettes were a big thing, regarded as a symbol of success and class. The problem was that the powder forming the "gold" tip would come off onto smokers' lips. Pfleumer addressed this by embedding metal particles in a plastic binder.

After that, he realized a similar approach could be taken as an alternative to music recording. A big fan of the Dresden Opera, he wanted to find a way to improve on the rasping, scratchy phonographs of the day. He began to wonder if there was any reason why he should not coat a magnetic strip in a similar way to the gold tips of those cigarettes.

He began to experiment with a range of materials. In 1927, he made it work. Using very thin paper, he coated it with iron oxide powder and used lacquer as adhesive. He applied for a patent for the technique, which was granted in 1928. Recorded music and speech would never be the same again.

Four years later, the project began to become commercial. In December 1932, Pfleumer granted AEG, a large electrical company in Berlin, the right to use his invention to construct the world's first tape recorder. It was named the Magnetophon K1. It was unveiled to the public in 1935, when it was demonstrated at the IFA.

Then, the first tapes were shipped for successful use on recorders built by AEG. At first, there was minimal commercial interest in the format but then fate intervened. The fearsome Reich Marshal Hermann Goering recorded an epic speech on wax disks. When he lost part of the recording, he quickly embraced the tape format.

A new format was born, but Pfleumer died before it became a mass-market system for recording and playing music. He had settled in Dresden in 1945. Soon after the city was destroyed in the Second World War, he passed away.

BEST KNOWN TODAY FOR:
Inventing the cassette tape.

Alexander Fleming
(1881–1955)

In 1940, two scientists demonstrated that penicillin could be used as a drug to fight a number of bacterial diseases. Alexander Fleming was neither of these scientists, but without him, they would never have made their big breakthrough.

Born in Lochfield, in Ayrshire, Scotland on August 6, 1881, Fleming moved to London at the age of 13. He studied at Regent Street Polytechnic, worked as a shipping clerk, and served in the army during the Boer War.

He went on to train as a doctor and qualified with distinction in 1906. Working under Sir Almroth Wright, a pioneer in vaccine therapy, Fleming undertook research at St Mary's Hospital Medical School at the University of London until his work was interrupted by the outbreak of the First World War. He served with the Army Medical Corps at a wound-research laboratory in Boulogne, France. His endeavors there were mentioned in dispatches.

Back at St Marys in 1928, he began to study influenza and made a world-changing discovery, quite by accident. Returning to the laboratory after a holiday, he saw that a blob of mold had grown on a dirty dish he had absentmindedly left out. He noticed, though, that no bacteria were growing around the blob. Fleming concluded that something in the mold was killing the bacteria. He discovered the active substance, which he named penicillin. "When I woke up just after dawn on September 28, 1928, I certainly didn't plan to revolutionize all medicine by discovering the world's first antibiotic, or bacteria killer," he said later. "But I suppose that was exactly what I did."

It was exactly what he did, because some years later, Australian Howard Florey and Ernst Chain, a refugee from Nazi Germany, developed penicillin further so that it could be used as a drug. By the 1940s it was being mass-produced by the American drugs industry.

Fleming wrote influential papers on bacteriology, immunology, and chemotherapy. He was elected fellow of the Royal Society in 1943 and knighted in 1944. In 1945 he, Florey and Chain shared the Nobel Prize in Medicine. In 1999, he was named in *Time* magazine's list of the "100 Most Important People" of the twentieth century.

On March 11, 1955, he died of a heart attack at his London home. He was buried in St Paul's Cathedral. He had saved millions of lives.

BEST KNOWN TODAY FOR:
Discovering penicillin, kick-starting the age of modern antibiotics.

Emmy Noether
(1882–1935)

Albert Einstein hailed her as the most "significant" and "creative" female mathematician of all time. Some argue that she was the greatest mathematician, period. Many contend that her theorem is as important as Einstein's theory of relativity.

But don't be surprised if you've never heard of Emmy Noether. As a female, Jewish mathematician in the early decades of the twentieth century she faced plenty of forces that were keen to erase her contribution.

Noether was born in Erlangen, Germany. Her father was a distinguished math professor and her brother Fritz was an applied mathematician. The algebraic blood of her family veins served her well, despite the obstacles of the day.

Although she studied mathematics at the University of Erlangen, she was only allowed to audit classes rather than participate fully because of her sex. When she taught at the University of Göttingen, her lectures were often advertised under the name of one of her male colleagues.

Noether's primary field remained mathematics, but the abstract work she was doing proved to have huge consequences for the world of physics. She developed a theorem that states that every differentiable symmetry of the action of a physical system has a corresponding conservation law. It means, for example, that momentum will remain constant as long as the physical system remains symmetrical. Noether's theorem is now part of the bedrock of physics and work like the hunt for the Higgs boson particle, the existence which was theorized on premises established by the theorem, would never have been possible without it.

With the rise of the Nazis, Noether fled Germany and, with the patronage of Einstein, was given a job at Bryn Mawr College in Pennsylvania, America. Only 18 months after her arrival in the United States, at the age of 53, Noether was operated on for an ovarian cyst, and died within days.

Throughout her life Noether went on to gain the admiration of her peers and students, but she struggled to gain paid roles and often had to rely on the championing of her male colleagues. The papers she wrote in the fields of abstract algebra and ring theory were sometimes published under a male pseudonym. Even today she is barely known, even in circles whose work relies upon her findings.

BEST KNOWN TODAY FOR:
Her theorem that, by linking geometry in nature and the forces that describe its behavior, gave thinkers a new tool to understand the universe and underpins much modern thinking in physics.

Percy Shaw
(1890–1976)

Rites of passage for children come when they discover that an assumption they held was not true. It can be a shock when they realize that Father Christmas and the tooth fairy are not real, for instance.

Another awakening can come when they realize that the cats' eyes they see in the road at night do not come from actual cats.

Their creator, Percy Shaw, was born in Halifax in the West Riding of Yorkshire. His was a large family: his parents had four children together and his father also had seven from his first marriage. Shaw was inventive as a child, a streak that was bolstered by a strong work ethic, and eventually paid off. He worked in a local mill when he was 13, and during the First World War he started a business that repaired small machine tools. He also made money by selling vegetables from the family's garden.

Later, he produced perhaps the best invention in the history of road safety.

Inspiration came to him on a dark and foggy night in 1933 as Shaw was driving down a steep winding road to his home in Boothtown. During previous nighttime journeys along this road, he had negotiated the dangerous bends by using the reflections of his car headlights on the tramlines.

However, on this fateful night, the tramlines had been removed for repair.

Shaw was suddenly plunged into pitch darkness. But then, he saw two points of light ahead. The reason? His headlights had caught the eyes of a cat, which was sitting on a fence.

It was a eureka moment: Shaw realized he could revolutionize road safety by creating a simple reflecting device that could be fitted to all road surfaces. Drivers could then identify the division between lanes and the edges of the roads at night.

And what better form could this take than a pair of small studs, much like the cat's eyes that had saved him? He set to work on trials and took out patents on his invention in April 1934. The following year he formed a company, Reflecting Roadstuds Ltd, with himself as Managing Director.

It was slow going at first, but before long, the company was making over a million cats' eyes a year, helped along by the blackouts during the second world war. After the war, he got the backing of a Ministry of Transport committee led by James Callaghan. Then, their use spread across the globe. Roads had never been so safe at night.

BEST KNOWN TODAY FOR:
Making our roads safer at night by inventing cats' eyes.

John von Neumann
(1903–1957)

In his heyday, John von Neumann was regarded
as the smartest man on the planet.
This would be a fine achievement at any point
in human history, but during the era of
Albert Einstein it was a mighty feat.

Born in 1903, it soon became clear that von Neumann had no ordinary mind. At the age of eight, he could divide two eight-digit numbers in his head. For a bit of fun, his parents would challenge him to rapidly memorize a page from a telephone book and listen with wonder as he accurately recited its numbers and addresses.

He emigrated to the United States in 1930 where he became a seminal thinker on mathematics, physics, and computers. He joined the Manhattan Project, which developed the world's first atomic bomb.

During the 1940s, he paved the way for the computer as we know it when he created a form of programming called the von Neumann architecture. He demonstrated for the first time that a computer could be a general-purpose machine—that it could have a fixed structure but still perform different tasks without the need to change its physical circuitry.

He also formulated game theory, which entails devising "games" to simulate situations of conflict or cooperation. This method of logical thinking allows researchers to unravel decision-making strategies, and understand why certain types of behavior occur.

Originally applied to "zero-sum games" dealing with the balance of loss and gain between "players," the theory was obviously useful in the field of economics. However, it also gave him the means to approach the Cold War strategy of Mutually Assured Destruction. Having helped create the atom bomb, he was now allowing humankind to understand its geopolitical consequences.

Game theory has since been adopted in evolutionary biology, computing, and philosophy as a means of helping us to understand the science of logical decision-making in humans, animals, and computers.

Von Neumann could be a fierce character. At Princeton University he blasted German marching music on his office gramophone. He reportedly once told rival mathematician John Nash that his work on the concept of equilibrium was "trivial."

The debate continues as to who had the greater mind: von Neumann or Einstein. Eugene Wigner, who knew both, said, "No one had a mind as quick and acute as John von Neumann."

BEST KNOWN TODAY FOR:
Game theory, his tool for logical decision-making, which has been widely taken up and developed across many fields.

J. Robert Oppenheimer
(1904–1967)

Looking back over his formative years, J. Robert Oppenheimer remembered himself as an "unctuous, repulsively good little boy" whose comfy New York upbringing left him ill-prepared for the "cruel and bitter things" of life. He was never taught, he said, a "normal, healthy way to be a bastard."

Born in 1904, by the age of 10 he was already studying advanced concepts in minerals, physics, and chemistry. Then he began writing letters to the New York Mineralogical Club, which duly invited him to deliver a lecture—not realizing that their clever correspondent was a twelve-year-old boy.

He went on to study physics at Cambridge, Harvard, and Berkeley and was subsequently chosen as scientific director of the Manhattan Project, the quest to build an atomic bomb for President Roosevelt.

Despite his lack of experience, Oppenheimer excelled not only as the scientific brains behind the bomb, but he also became quite the manager, deftly harnessing a squad of cranky geniuses who had been thrown together to build a bomb that could kill millions.

The project reached its climax on July 16, 1945, in New Mexico, where the first atomic bomb was detonated. As he watched the mushroom cloud rise, Oppenheimer was reminded of the Hindu god Krishna's words in the Bhagavad Gita: "Now I am become Death, the destroyer of worlds."

However, a colleague, the physicist Isidor Rabi, remembers him having a more victorious air. "His walk was like High Noon … this kind of strut," claimed Rabi. Certainly, Oppenheimer was clear about what was at stake. "We knew the world would not be the same," he later recalled.

After learning of the carnage when atom bombs were dropped on Hiroshima and Nagasaki, Oppenheimer became appalled at what he had unleashed. When he told Harry Truman that he felt he had blood on his hands, the president told his aides he never wanted to see this "crybaby scientist" again.

Oppenheimer tried to halt the nuclear arms race, and he opposed the creation of the hydrogen bomb. The authorities were determined to shut him up and, using FBI files detailing his contact with communists in the 1930s, they cast him as a Soviet spy.

An unjustly discredited man, he retreated to Princeton, although he continued to lecture, write, and work on physics. He died of throat cancer in 1967.

BEST KNOWN TODAY FOR:
Bringing the atomic bomb into the world.

Grace Hopper
(1906–1992)

If you wanted to get on the wrong side of Grace Hopper, there was always one reliable way of doing it. All you had to do was to respond to her innovative suggestions by saying: "But we've always done it this way." For her, this mentality was infuriating.

Hopper was born in New York in 1906 and showed a flair for mathematics at school. She graduated from Vassar College in 1928 with a mathematics degree, then received Master's and Doctorate degrees in mathematics and physics from Yale University.

She enlisted as a Navy Reserve during the Second World War when she was 37 years old. In 1944, she was assigned to the Bureau of Ordnance Computation Project at Harvard University. There, she worked as a programmer on a calculating device called the Mark I, which was a precursor of electronic computers.

Her finest hour came in 1949, when she helped to create the first all-electronic digital computer. It was named the Universal Automatic Computer (UNIVAC).

Hopper also helped make coding languages more practical and understandable, and she created the first computer compiler, which translated source code from one language into another. It opened the way for the distinction between hardware and software, something we now take for granted.

In 1966 she retired from the Reserves, but when she returned for active duty in 1967 for what was supposed to be a brief, six-month spell, she ended up staying on for a further 19 years. In fact, Hopper only put her feet up following her (non-voluntary) retirement in 1986 at the age of 79.

Her mind was mathematical but also witty and feisty. She coined the term "bug" for computer failures after one of the early computers she worked on broke down because of a two-inch moth that had become embedded in it. From thereon in, whenever a computer developed a mysterious problem, they would say: "It must have a bug in it."

She was posthumously awarded the Presidential Medal of Freedom by President Barack Obama during a White House ceremony in 2016. "If Wright is flight, and Edison is light, then Hopper is code," said Obama.

This pioneer of computer programming was often referred to as the "first lady of software." Thanks to her remarkable achievements, she was also known to many, quite accurately, as Amazing Grace.

BEST KNOWN TODAY FOR:
Pioneering computer programming, and coining the term "bug" to describe a fault with our computers.

Alan Turing

(1912–1954)

One of his school reports once stated that Alan Turing needed to think again if he wanted to get into Cambridge University. The report, from Sherborne School in Dorset, England, warned: "He must remember that Cambridge will want sound knowledge rather than vague ideas."

Born in 1912 in Maida Vale, London, England, Turing was indeed accepted to Cambridge, and studied there from 1931 to 1934. He ran far ahead of the syllabus, in his dissertation he proved the central limit theorem, and he was elected a fellow after graduation.

In 1936, he penned his paper "On Computable Numbers," which is now recognized as the foundation of computer science. He invented the idea of a "Universal Machine" that could decode and perform any set of instructions. It was later named the Turing machine and is considered a model of a general-purpose computer.

After a spell in America, Turing returned to Britain in 1939. He took up a full-time role at Bletchley Park in Buckinghamshire cracking the "Enigma" code. The Enigma was a type of enciphering machine used by the German armed forces to send messages securely during the Second World War. His contribution helped the Allies to defeat the Nazis and win the war.

After the war he worked at the National Physical Laboratory, where he devised one of the first designs for a stored-program computer and a project sometimes described as an "electronic brain." He also joined the Victoria University of Manchester, where he helped develop the Manchester computers.

Winston Churchill said Turing made the single biggest contribution to Allied victory, so he should have been a national hero. Instead, in 1952 he was prosecuted for homosexual acts, which were still illegal in the UK. He accepted chemical castration treatment as an alternative to prison.

He was found dead in bed by his cleaner on June 8, 1954. He had died from cyanide poisoning the day before. A partly eaten apple lay next to his body; the coroner's verdict was suicide.

In 2009, following a popular online campaign, UK Prime Minister Gordon Brown made an official public apology on behalf of the British government for "the appalling way Turing was treated." Queen Elizabeth II granted him a posthumous pardon in 2013.

Turing felt that life was full of challenge and opportunity. He said: "We can only see a short distance ahead, but we can see plenty there that needs to be done."

BEST KNOWN TODAY FOR:
Cracking the enigma code to help the Allies to victory over the Nazis, while the "Turing test" means of distinguishing AI behavior from human behavior is referenced in everything from popular films to the "CAPTCHA" test used on websites today.

ALAN TURING

Hedy Lamarr

(1914–2000)

Becoming the first actress to depict an orgasm on screen is quite an achievement. Inventing the technology that paves the way for the mobile phone and wifi connection is also something of a feather in the cap. Managing to do both is a life well lived.

Born in 1914, Hedwig Kiesler grew up in Vienna. Her father taught her engineering, but the teenage Kiesler had other ideas—she dreamed of becoming an actress.

One day, at the age of 15, she played truant from school and presented herself to the headquarters of a film studio. She charmed them and they took her on immediately as a script clerk. Before long she was working as a supporting actress.

She impressed: at the age of 18 she took the lead role in Gustav Machaty's *Ecstasy*, in which she performed her groundbreaking and controversial climax. She explained later that to make her cries seem authentic, her male co-star stabbed her with a safety pin.

She became a celebrity of her day. Described as a "porcelain-skinned, raven-haired goddess," she married and divorced six times, appeared in a series of films and thrilled audiences with her glamorous accent. Unlike most celebrities, however, she was also an inventor. She often eschewed Hollywood parties, preferring to stay at home and experiment with soluble tablets, or tweak her design for a traffic light.

During the Second World War, she devised a radio-guided torpedo system, which led to a "spread-spectrum" technology. She hoped this could be used by the Allies to stop enemies jamming the radio signals, but her invention was rejected by the US Navy.

She retreated to Florida, and life became less happy. She had disastrous plastic surgery, got arrested for shoplifting, and took her publisher to court in an unsuccessful bid to block the release of her own salacious, ghostwritten memoir.

It was only decades on from her spread-spectrum innovation that she gained recognition for it when it was incorporated into mobile-phone technology, wifi, and Bluetooth. Then, in her eighties, the Electronic Frontier Foundation honored her with its Pioneer Award. She was also inducted into the National Inventors Hull of Fame.

She died in Altamonte Springs, Florida in January 2000 at the age of 85. Her death certificate cited three causes: heart failure, chronic valvular heart disease, and arteriosclerotic heart disease. She had never been one to do things by halves.

BEST KNOWN TODAY FOR:
Being both one of the most beautiful actresses of her era, and an inventor to whom bluetooth and wifi technology owe a debt of influence.

CHAPTER

4

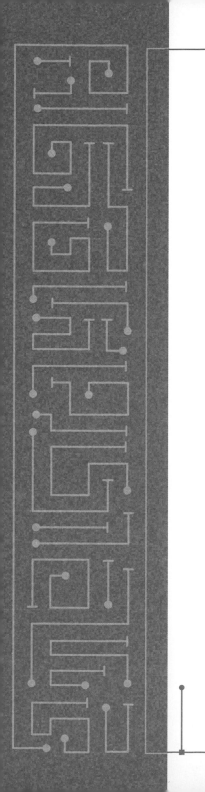

SPACE-AGE SAGES

Katherine Johnson

Gene Roddenberry

Kathleen McNulty

Marie van Brittan Brown

Douglas Engelbart

Robert Noyce

Vera Rubin

Marty Cooper

Yuri Gagarin

Jane Goodall

Richard Saul Wurman

Harrison Schmitt

Katherine Johnson

(1918–)

Born in White Sulphur Springs, West Virginia, in 1918, her way with numbers catapulted Katherine Johnson several grades ahead in school. In 1939, she was chosen as one of three black students to integrate West Virginia's graduate schools.

After a spell as a teacher, in 1952 she took a position at the all-black West Area Computing section at the National Advisory Committee for Aeronautics' Langley laboratory. One of her first tasks was working on the investigation of a plane crash caused by wake turbulence. Then she helped plan the mission that saw an American astronaut orbit the Earth for the first time.

Complex mathematics was involved in the orbital equations that would control the trajectory of the capsule for John Glenn's mission in 1962. Glenn's life depended on the calculations being correct. As he was checking the pre-flight checklist he told the engineers to "get the girl" to double-check the calculations. "If she says they're good," Johnson remembered Glenn saying, "then I'm ready to go."

He was indeed—he orbited the Earth three times, besting the achievements of Yuri Gagarin, who in 1961 had become the first human to enter outer space. Johnson went on to provide the calculations for NASA that helped synchronize Project Apollo's Lunar Lander with the moon-orbiting Command and Service Module, and worked on the Space Shuttle and the Earth Resources Satellite.

Yet even as Johnson had been the human computer behind such a moment of national pride, she was facing huge inequality. The Civil Rights Act, which ended local and state segregation, did not come into force until 1964. She also suffered from sexism: it was widely assumed women couldn't handle mathematics as well as men.

In 2016, a movie called *Hidden Figures* (based on a non-fiction book of the same name, also released in 2016) finally gave recognition to Johnson and the other black female mathematicians who worked at NASA, Dorothy Vaughan and Mary Jackson.

Johnson was a trailblazer who broke down barriers throughout her life. As such, she became an icon and inspiration to many and was justly celebrated when, in 2015, she received the Presidential Medal of Freedom from President Obama. Yet, when the Katherine G. Johnson Computational Research Facility in Hampton, Virginia, was named after her, she wondered what all the fuss was about. "You want my honest answer? I think they're crazy," she said.

BEST KNOWN TODAY FOR:
Overcoming racial prejudice and institutional sexism to provide the raw data that made early NASA space missions possible.

Gene Roddenberry
(1921–1991)

Gene Roddenberry followed in the footsteps of his father when he joined the Los Angeles Police Department, but his time there would ultimately take him out of this world and into "space, the final frontier."

After serving in the traffic department for a while, he became a speechwriter for the Chief of Police. This led to him taking up a new side project: writing for television. He penned scripts for the crime drama *Highway Patrol* and the Western series *Have Gun—Will Travel*. Then he created his own serial: a Cold-War saga entitled *The Lieutenant*.

His big break came in 1964, when he created *Star Trek*. The series followed the interstellar adventures of Captain James T. Kirk and his crew aboard the Starship USS Enterprise in the twenty-third century. It ran for three seasons and then spawned big-screen feature films, which Roddenberry was involved in as a producer.

Following a costly divorce in 1969, the writer supported himself by appearing at sci-fi conventions. As well as bolstering his bank balance, these slots also created a snowball of support for *Star Trek* to return.

In 1972, *TV Guide* described Roddenberry's creation as "the show that won't die"—and it was. Viewers loved to leave behind their worries and hop aboard the USS Enterprise to "explore strange new worlds, to seek out new life and new civilizations, to boldly go where no man has gone before." The series captured something of the zeitgeist, portraying an ideal vision of a peaceful and tolerant future human civilization.

Finally, in 1987 the franchise returned to the small screen, in the form of *Star Trek: The Next Generation*. Although Roddenberry was less involved than previously, he remained a consultant on the series, which retained its iconic status.

He died in 1991 at the age of 70, but his story continues beyond his death. In 1992, he was the subject of the first space "burial" when the NASA space shuttle Columbia carried a portion of his cremated remains into space and returned them to Earth.

The influence of sci-fi culture should not be underestimated. For Rodenberry, his obsession with the genre began at school, when a classmate lent him a copy of *Astounding Stories*. For Trekkies across the planet, his own sci-fi creation, with its tone of progressive optimism, would prove just as epochal. Not least for George Lucas, who said that *Star Wars* would not have been possible without *Star Trek*.

BEST KNOWN TODAY FOR:
Boldly going where no man has gone before. That is, the creation of *Star Trek*.

Kathleen McNulty

(1921–2006)

Wars shaped the life of Kathleen McNulty. She was born on February 12, 1921 in Creeslough, Ireland, as the Irish War of Independence was being fought. After her Republican father was arrested and released, when she was three, the family moved to Pennsylvania, in the United States. It was quite an upheaval for her: she was unable to speak any English.

McNulty eventually settled and became a mathematics whiz at school, excelling at algebra, plane geometry, trigonometry, and solid geometry. In 1942, she earned a degree in maths from Chestnut Hill College for Women. Only three women in her class of 92 graduated in mathematics. After graduation, she took a role at the US Army's Ballistics Research Laboratory in Maryland. The Second World War was in full flow, and her first role involved calculating trajectories for shells and bullets. This was life-saving information for soldiers using artillery guns in battle.

However, it soon became clear that analog machines were not quick enough for the task, and that an electronic rather than a human computer would be needed to keep up. Engineers created the world's first general purpose digital computer, called ENIAC—Electronic Numerical Integrator and Computer. McNulty became one of the first programmers to work on it.

She and five other women—Jean Jennings Bartik, Frances "Betty" Snyder Holberton, Marlyn Wescoff Meltzer, Frances Bilas Spence, and Ruth Lichterman Teitelbaum— taught themselves how to perform calculations in 15 seconds. It was a focused and professional effort. "None of us girls were ever introduced," said Kathleen; "We were just programmers." Her colleague Meltzer said: "We were sure this machine could do anything we wanted it to do. We were very cocky about that."

These brilliant young women had learned to program without programming languages or tools, just logical diagrams. But when ENIAC was revealed to the public in 1946, their contribution was not publicly acknowledged.

Even after she married in 1948 and became a full-time housewife and mother to seven children, McNulty continued to work on computer program designs and techniques. In time, recognition came her way and she was alive to see some of it. In 1997, she was inducted into the Women in Technology International Hull of Fame. In July 2017, Dublin City University named its computing building after her.

BEST KNOWN TODAY FOR:
Being one of the long-unacknowledged female geniuses who programmed the first general-purpose electronic digital computer.

Marie van Brittan Brown

(1922–1999)

They say necessity is
the mother of invention—
anxiety can be, too.

During the 1960s, nurse Marie van Brittan Brown often worked long, late hours. Her electrician husband was sometimes away when she returned to Jamaica, Queens in New York City, so she felt uneasy when alone in their home. Crime in the neighborhood was high and police responses were notoriously slow.

So she took matters into her own hands and designed a closed-circuit security system that monitored visitors with a camera and broadcast their images onto a television monitor in her bedroom and other parts of the residence. If the visitor was a friend, she could press a button to allow them entry. But if they looked like trouble, she had a panic button that contacted the police.

It turned into an ingenious concept that proved she was ahead of her time. The camera would journey past a series of four peepholes at various heights, so visitors of any size could be identified and spoken with. There was also a radio-controlled wireless system to feed the images seen at the door back to the monitor.

Brown had invented the home security system and brought CCTV, previously the preserve of military surveillance, into the domestic home. Units based on her invention are used in flats, homes, and workplaces across the globe.

Along with her husband, she was awarded a patent on December 2, 1969.

Speaking to the *New York Times* a few days later, she said: "A woman alone could set off an alarm immediately by pressing a button, or if the system were installed in a doctor's office, it might prevent holdups by drug addicts."

Her patent has since been referenced by 13 inventors who say they can link their own concepts back to Marie's closed-circuit system. She received an award from the National Scientists Committee. As of 2016, some 100 million concealed closed-circuit cameras were in operation worldwide.

BEST KNOWN TODAY FOR:
Inventing the home security system and bringing CCTV into the home.

Douglas Engelbart

(1925–2013)

He was barely into his twenties, but Douglas Engelbart had had enough of the mainstream way of living. He decided that rather than having a "steady job," he would devote his time and energy into making the world a better place.

Engelbart reasoned that improving the world required a collective effort, one that harnessed the brains of all people. He further concluded that computers, which were used rarely at this point, would be the vehicle for taking this forward.

Born in Portland, Oregon, in 1925, he served two years as a US Navy radar technician in the Philippines. During this spell, he read "As We May Think," a visionary 1945 essay by Vannevar Bush that anticipated several features of information society. The key part of that society for Engelbart would become the computer. When you sit at your computer in the twenty-first century, Engelbart's legacy is right there in front of you. In 1967, he applied for the patent for a wooden shell resting on two small metal wheels, which would be used as a "position indicator for a display system."

Engelbart and his team came up with two nicknames for the device. First, they called it a "bug." Then, noticing that the wire coming out of the end of it looked like a tail, they came up with the name that stuck: the "mouse."

Before you ask, Engelbart never received so much as a dime in royalties for its invention. He says that although his Stanford research institute patented the mouse, it had no idea of its value. "Some years later it was learned that they had licenced it to Apple Computer for something like $40,000," he explained. In 2017, a single Logitech Air 3D Laser Mouse in Gold Case was on sale for more than half that sum.

Engelbart's other standout achievement was the observation that the intrinsic rate of human performance is exponential. We can use technology all we like, he realized, but the ability to improve on our improvements rests entirely with humanity. This became known as Engelbart's Law. But it was his invention of the computer mouse during the summer of love that clicks most regularly in the present day.

BEST KNOWN TODAY FOR:
Launching an infinity of clicks by inventing the computer mouse.

Robert Noyce
(1927–1990)

"Optimism is an essential ingredient of innovation," said Robert Noyce. "How else can the individual welcome change over security, adventure over staying in safe places?"

Known by a variety of nicknames including "the mayor of Silicon Valley" and "the podfather," he was quite the innovator.

Born on December 12, 1927 in Grinnell, Iowa, the son of a preacher man, he was a feisty child. At the age of five, he beat his father at ping-pong. When his mother said it was nice of daddy to let him win, Robert angrily corrected her. "That's not the game!" he said; "If you're going to play, play to win!"

His childhood triumphs included building a radio from scratch and constructing a child-sized aircraft, which he flew from the roof of his home.

After studying at the Massachusetts Institute of Technology, in 1957 he founded the company Fairchild Semiconductor, where he devised the idea of the integrated circuit—a chip of silicon with many transistors etched into it. This is now better known as the microchip. At a stroke, he had sparked the personal-computer revolution.

Then, in 1968, along with his friend Gordon Moore, he founded Intel, which became responsible for more than 80 percent of the microprocessors in personal computers. Dozens of technology companies stemmed from his firms. So did a culture: the atmosphere of informal, maverick innovation at Silicon Valley originated with Noyce. He believed in an unconventional management style and eschewed executive perks such as flash company cars and designated parking spaces.

Instead, he built a reputation as an inspiring visionary during his reign as one of the first scientists to work in the stretch of California now known as Silicon Valley. Away from work he was just as energized: he sang madrigals, read the novels of Ernest Hemingway, flew his own airplane and scuba dived.

During the 1980s, President Ronald Reagan awarded him the National Medal of Technology, and he was inducted into the US Business Hull of Fame. On December 12, 2011, a Google Doodle celebrated the 84th anniversary of his birth.

The author Tom Wolfe said Noyce had a defining characteristic: the halo effect. "People with the halo effect seem to know exactly what they're doing and moreover make you want to admire them for it," he said. "They make you see the halos over their heads."

BEST KNOWN TODAY FOR:
Co-founding Intel and forging the Silicon Valley culture in his maverick image.

ROBERT NOYCE

Vera Rubin
(1928–2016)

Her pioneering work contributed to the theory of dark matter but she faced a lifetime of obstacles and cynicism, and was denied a Nobel Prize. For Vera Rubin, though, her efforts had always been about the joy of discovery, rather than the glow of recognition.

Rubin became fascinated by astronomy as a young girl. At the age of 10, she gazed at the night-time sky from her north-facing bedroom in Washington D.C. Her father encouraged her burgeoning interest: he helped her build a telescope and took her to meetings of amateur astronomers.

She was the only astronomy major to graduate from the prestigious women's college Vassar in 1948. She applied to be a graduate student at Princeton but was told that women were not allowed in the university's graduate astronomy program. That policy was not abandoned until 1975—but by then she had already changed the way humankind looks at the universe.

After being rejected by Princeton, Rubin applied to Cornell University, where she was accepted to study physics. She then earned a doctorate at Georgetown University, and took a research position at the Carnegie Institution in Washington, where she focused on the dynamics of galaxies.

During the 1970s, she made a historic breakthrough when she helped prove that the stars at the edges of galaxies moved faster than expected. Gravity calculations using only visible matter in galaxies showed that the outer stars should have been moving more slowly. What was the cause of the discrepancy? For decades, astronomers had theorized that there was hypothetical matter we cannot see, called "dark matter." Rubin's discovery seemed to prove its existence. At first, astronomers were reluctant to accept her conclusion; however, the evidence she provided would be confirmed as accurate and the naysayers would be forced to agree that she had been right all along.

Rubin continued her career of incredible discoveries. In 1992, she discovered a galaxy in which half the stars in the disc orbited in one direction and half in the opposite direction, with both systems intermingled. It was named NGC 4550.

Although she devoted her life to solving the mysteries of the universe, Rubin was charmed by the mystique of it all. "We have peered into a new world," she once wrote, "and have seen that it is more mysterious and more complex than we had imagined."

Thrillingly, she added: "Still more mysteries of the universe remain hidden. Their discovery awaits the adventurous scientists of the future. I like it this way."

BEST KNOWN TODAY FOR:
Finding evidence of dark matter.

Marty Cooper

(1928–)

Mobile, or cell, phones have become so sophisticated, and such a fundamental part of our lives, that it's easy to forget what a recent invention they are. The man we can thank for them is Marty Cooper.

Born in Chicago to Ukrainian immigrants, he graduated from Illinois Institute of Technology in 1950 and served in the navy during the Korean War. In 1954, he joined Motorola as a senior development engineer. There, he produced the first cellular-like portable handheld radio system. It was snapped-up by the Chicago police department in 1967. As the chief of Motorola's communications systems division, in 1973 he conceived and developed the first portable cellular phone and began the 10-year process of bringing it to market.

The idea of a personal telephone, which you could use anywhere, was astonishingly radical at the time. Even though the public was excited by the prospect of it, it took 20 years and a $100 million investment until Motorola turned a profit on the project.

The original model weighed 2.5 pounds (1.1 kg), was 10 inches (25 cm) long, and was nicknamed "the brick." It offered just 20 minutes of talk time before requiring a 10-hour recharge—but, as Cooper pointed out, this was not a particular issue as no one would be able to hold the phone up for much longer.

For the first call, Cooper stood on a New York street and dialed a man at a rival telecoms company and announced he was speaking from "a real cell phone." In the era of Twitter trolls glued to their handsets, it seems somehow fitting that the first mobile phone conversation was a wind-up.

Cooper also made an astonishing discovery: that the maximum number of conversations (voice or data) that can be conducted in all of the useful radio spectrum has doubled every two-and-a-half years for the past 104 years. This observation became known as Cooper's Law.

Cooper continued to sit on the boards of companies and on government committees long past the retirement age. He is the co-founder of several communications firms with his wife Arlene Harris, who is known as the "first lady of wireless."

Even though he remains fascinated by the device, trying out a different model phone every couple of months, his verdict on the unintuitive, feature-heavy modern mobile phone is damning. They are, according to Cooper, a "monstrosity."

BEST KNOWN TODAY FOR:
Putting the cell/mobile phone in our hands.

MARTY COOPER

Yuri Gagarin
(1934–1968)

During the Second World War, the teenage Yuri Gagarin watched in wonder as a Soviet Yak fighter plane made an emergency landing near his home in Klushino. Two decades later he made a dramatic touchdown of his own, when he emerged from a spaceship that had landed near the village of Smelovka.

His aeronautical career began as a teenager when he took on weekend training as a Soviet air cadet at a local flying club. When he was drafted by the Soviet Army in the 1950s, he was sent to the First Chkalov Air Force Pilot's School in Orenburg. By the end of the decade, he was a lieutenant.

The 1960s saw Gagarin join the Soviet space program. He was selected for the debut space mission after a vigorous selection process and more hurdles than most Olympians could manage. The tests included a psychological assessment, which characterized Gagarin as "modest," with a "high degree of intellectual development," and concluded that he "handles celestial mechanics and mathematical formulae with ease as well as excels in higher mathematics."

Gagarin was 27 years old when, aboard the spacecraft Vostok 1, he became the first human being to travel into space. In this spacecraft he completed an orbit of the planet in 89 minutes at a maximum altitude of 300 km (187 miles).

He was in space for a total of one hour and 48 minutes. The only statement he was recorded as making was, "Flight is proceeding normally; I am well." For such galactic circumstances this had a remarkably down-to-earth attitude toward it.

Physically, he came back to Earth with a bang. A spherical capsule, the Vostok was designed to eliminate changes in center of gravity so it could stay comfy for its one-man crew no matter its orientation. However, unlike latter Soviet spaceships, it was not fitted with thrusters to help slow it down as it headed back toward Earth, meaning he had to eject before reaching ground.

Russia's breakthrough was a Cold War triumph. Gagarin himself did fairly well out of it, too. He became a celebrity, and was showered with honors and awards, including "Hero of the Soviet Union," the nation's greatest gong. Klushino, the village he grew up in, was renamed Gagarin after his tragic death in 1968, when his MiG-15 fighter jet crashed during a training exercise.

BEST KNOWN TODAY FOR:
Being the first man in space.

Jane Goodall

(1934–)

On her first birthday, in 1935, Jane Goodall's parents presented their baby girl with a soft-toy replica of a chimpanzee. It might just have been the most prescient present ever given to a baby.

Goodall's subsequent study of chimpanzees in Tanzania has revolutionized our understanding of the primates. Her new insights into our closest living relatives also challenged fundamental assumptions of what it means to be human.

She had dreamed of Africa from her early years and developed strong ideas about the place. As a child, when she was taken to the cinema to see a Tarzan film, she burst into tears at the sight of the lead actor, Johnny Weissmuller. In the foyer, she told her mother, "That is not Tarzan." When she finally got to Africa, it lived up to her expectations. She was thrilled by the lions, rhinos, and giraffes. "I often think that's one of the most magical times of my whole life," she recalled.

As she lived among wild chimps in Tanzania, she observed them using sticks and grasses as tools. Until then, it was thought that only humans used and made tools. When she told her boss what she had witnessed, he understood the weight of the moment. "Now we must redefine man, redefine tools, or accept chimpanzees as humans," he said. Redefinitions became even more urgent when Goodall noticed that chimps had distinct personalities and that they embraced, kissed, and tickled each other. She also pointed out that the primates were capable of political maneuvering to get what they wanted, and of waging war on one another.

She took from all this that many human behaviors may have been inherited from the common ancestors that Homo sapiens shared with chimpanzees six million years ago. It has always been believed that these patterns of behavior were uniquely human. But, it turned out, humans aren't all that.

Now into her eighties, she has published a book of her findings, received nearly 50 honorary degrees, become a UN Messenger of Peace in 2002, and knighted Dame Jane in 2004. Through it all, however, she has remained utterly devoted to chimpanzees. She is also a tireless campaigner, pushing for people to develop a more enlightened attitude towards animals and climate change.

BEST KNOWN TODAY FOR:
Her work with chimpanzees that challenged our perception of higher primates—and our understanding of what it means to be human.

Richard Saul Wurman

(1935-)

For Richard Saul Wurman, "every book, every conference, every design" is simply an experiment on how he can get people "telling the truth." In the 1980s, he brought this philosophy to life when he launched what has become an iconic intellectual gathering.

Wurman, who graduated from the University of Pennsylvania in 1959, had noticed that technology, entertainment, and design were increasingly converging. So he mixed them together and formed a conference under the banner with the obvious acronym: TED.

What was his driving motivation? "I wanted to be around people who were smarter than me," he explained. His vision was for an informative yet informal gathering. "I wanted to get rid of the lectern, stay away from golf courses, not let people dress up because I don't own a suit," he explained.

The topics discussed at the first TED conference in 1984 date it better than anything: there was a demo of the compact disc, and one of the first presentations of the Apple Macintosh computer. Math genius Benoit Mandelbrot showed how to map coastlines using his developing theory of fractal geometry.

The event did not turn a profit and it was six years until the second TED conference was held. However, its 1990 comeback proved such a success that it became an annual event, initially held in Monterey, California. Within a few years it was a highlight of the intellectual calendar.

Since then, speakers have been drawn from wider backgrounds. As well as experts in technology, entertainment, and design, others to be invited include philanthropists, scientists, philosophers, musicians, business, and religious leaders.

Stephen Hawking, J. K. Rowling, and Bono are among those who have been called upon to deliver 18-minute lectures, all of which have been shared enthusiastically online. Speeches tend to be upbeat, inspiring, and lively, with pithy, attention-grabbing titles like "Information Is Food" and "Crowdsource Your Health." Sometimes speakers have gone a bit crazy. During a talk on malaria in 2009, Microsoft founder Bill Gates released live mosquitoes into the auditorium because "there's no reason why only poor people should have the experience."

In 2002, after 18 years, Wurman sold TED, but he continues to innovate. *Time* magazine describes him as an "old magician, in splendid, self-imposed exile." He has written, designed, and published 90 books.

BEST KNOWN TODAY FOR:
Launching the clever-clogs confab and multi-disciplinary conference for the viral age that is the TED talk.

Harrison Schmitt

(1935–)

Harrison Schmitt didn't grow up dreaming of flying to space, but when the opportunity arose to become a scientist-astronaut it was an easy decision.

"I thought about 10 seconds and raised my hand and volunteered," he recalled. He felt he had to, or he would end up regretting it when human beings actually did go to the moon.

Born on July 3, 1935, in Santa Rita, New Mexico, Schmitt went on to get a doctorate in geology from Harvard University in 1964. He was working with planetary geologist Eugene Shoemaker when he snapped up the role of a scientist-astronaut with NASA in June 1965.

At first, he trained moon-bound crews for their upcoming manned space flights. Crucially, he provided Apollo flight crews with expert instruction in lunar navigation and geology. He also pumped them with information on how to recognize and log geologic features on the Moon.

In December 1972, he got the chance to go to the Moon himself. As part of the Apollo 17 crew—the last mission to put men on the Moon—he set off to the Taurus-Littrow region, to discover whether it was a hotbed of volcanic activity. The crew consisted of commander Eugene Cernan, command module pilot Ron Evans, and Schmitt, who was the scientist and lunar module pilot.

Schmitt spent more than 20 hours on the surface and found the proof of volcanoes: orange volcanic beads that indicated a surface vent. "The orange soil never looks as orange to you in a picture as it did to us while we were on the moon," he recalled.

He surveyed the surface from his lunar rover, collected a record 113 kg (249 lb) of rock samples and took photographs. His love of Nordic skiing allowed him to glide effortlessly across the dusty lunar surface. "You use the same kind of rhythm, with a toe push," he explained.

A man of vast intelligence, he loved big-picture and abstract thinking but was also capable of plain common sense: when the fender of the lunar rover came off while they were on the Moon, he repaired it with duct tape.

Having become the first, and still the only, scientist to have ever walked on the Moon, Schmitt resigned from NASA in 1975. Two years later he turned to politics, embarking on a six-year term in the US Senate, representing New Mexico.

A remarkable life—the highlight of which were those 301 hours and 51 minutes he spent in outer space. "I recommend it," he says.

BEST KNOWN TODAY FOR:
Being the first scientist to walk on the moon and the most recent living human being to have done so, as part of the Apollo 17 mission.

HARRISON SCHMITT

CHAPTER

5

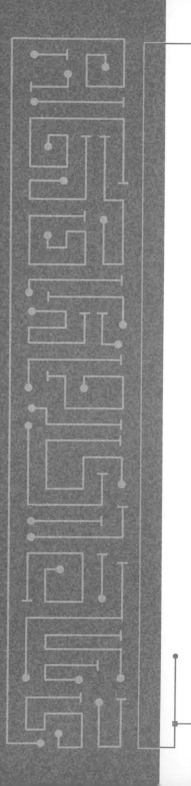

GEN-X GENIUS & MILLENNIAL MARVELS

Chuck Hull

Jerry Lawson

Stephen Hawking

Vint Cerf

Ward Christensen

Arsène Wenger

Steve Wozniak

Andy Hildebrand

Shigeru Miyamoto

Tim Berners-Lee

Bill Gates

David Braben

Rivers Cuomo

Larry Page

Limor Fried

Mark Zuckerberg

Chuck Hull

[1939–]

Chuck Hull's 3D printing came about because, while working in the furniture industry, he grew frustrated that the production of small plastic parts could take several months.

He wondered if he could form three-dimensional objects by piling thousands of thin layers of plastic on top of each other and then etching their shape using light. For years, he would spend his spare time in a small laboratory, trying to bring his vision to life.

He eventually refined a method in which light was shone into a vat of photopolymer to trace the shape of one level of the object. The other layers would then be printed until a finished object was produced. He phoned his wife, "got her out of her pajamas, told her to come down to the lab and see this."

He had just invented stereolithography—or 3D printing. In 1983, he created the first-ever 3D printed object: a small black eyewash cup. The following year he filed a patent for Stereolithography Apparatus.

In 1986, he co-founded 3D Systems to monetize the new method of production. It became the first 3D-printing company in the world. It sent its first printers, materials, and software to market in 1988.

Among its most enthusiastic early customers were car manufacturers, the aerospace sector, and companies designing medical equipment. General Motors and Mercedes-Benz used his technology to build prototypes. The medical industry uses it to create highly customized prosthetics, skin grafts, even new organs, and structural models in advance of surgery.

When he first invented 3D printing, Hull told his wife that it would take 30 years before his technology would be found in homes. His prediction proved spot on, and today the possibilities are huge

A sprightly and warm soul, when Hull retired he soon got bored and returned to the company he co-founded. He became Vice President as well as Chief Technology Officer, holding shares worth $20 million (£12 million).

In 2014, Chuck Hull was inducted into the National Inventors Hull of Fame and received the European Inventor Award. He has 93 patents to his name in the US and 20 in Europe. The 3D-printing industry is worth more than $3 billion annually.

BEST KNOWN TODAY FOR:
Being the father of 3D printing.

CHUCK HULL

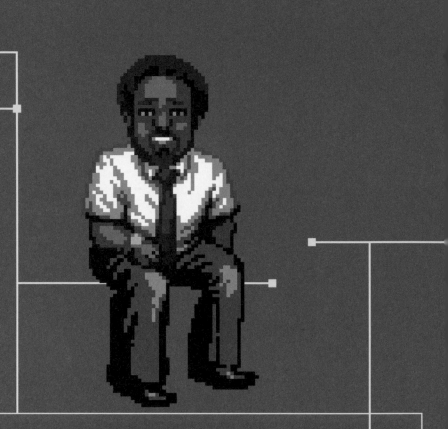

Jerry Lawson
(1940–2011)

Imagine that you were a computer geek in the 1970s. You crossed paths with Steve Jobs and Steve Wozniak shortly before they formed Apple, but you walked away, shrugging: "I was not impressed with them— either one of them, actually."

As you watched their ascent, you'd kick yourself, right? Well, not Gerald "Jerry" Lawson. He had his own trails to blaze.

Born in Brooklyn on December 1, 1940, he grew up in Queens with parents who supported his intellectual curiosity. So did his teachers: pointing to a picture of the inventor, George Washington Carver, one of them told Lawson, "This could be you." He was inspired.

"I want to be a scientist," he thought; "I want to be something." As a teenager he was crazy about chemistry, broadcast his own amateur radio station by running an antenna out of his window, and repaired his neighbors' television sets to earn money.

After attending both Queens College and City College of New York, he took a post as a roving design consultant at Fairchild in Silicon Valley. There, he invented an early coin-operated arcade game, entitled *Demolition Derby*.

But it was when he moved to the video game division that he made history. Lawson led the development of the Fairchild Channel F console, released in 1976, and then he and his team produced cartridges that could be loaded with different game programs and then inserted into the console one at a time.

Game programming had previously been soldered onto the game hardware. Now, players could buy a library of games—and manufacturers could sell them to them. For Channel F, the games included Space War, Blackjack and Bowling games. Lawson had paved the way for Atari, Nintendo, the Xbox, and Playstation.

He is one of the few African-American engineers to work in computing in the early days of the video game era. "I've had people look at me with total shock," he said. "Particularly if they hear my voice, because they think that all black people have a voice that sounds a certain way." He always made a point of he encouraging other young black men and women to become invested in science and engineering careers.

Lawson died in 2011 from complications with diabetes, a condition with which he'd been struggling for many years. On the eve of his death, he was honored as an industry pioneer by the International Game Developers Association.

BEST KNOWN TODAY FOR:
Inventing the video game cartridge—quite literally a game-changer in the industry.

Stephen Hawking
(1942–2018)

Among Stephen Hawking's many achievements, perhaps his most moving was the way he turned a crippling disease to his advantage. At the age of 21, when a form of motor neurone disease robbed him of the use of his limbs, he responded by intensifying his use of his brain.

Rather than tackling equations, he began to visualize problems in his mind to find a solution. It was an approach that would change the world.

At Cambridge University, as a postgraduate, Hawking began to study cosmology. Fascinated with how the universe began, he went on a hunt for a single theory that would describe everything. The theory he arrived at described how the universe was once concentrated in a single point, which then exploded in a "big bang." The big bang theory had been speculated about for several decades, but Hawking changed the whole story when he described it as being like the collapse of a black hole in reverse.

He published a paper showing that general relativity implies that the universe must have begun as a singularity. He broke further ground when he announced that black holes evaporate and shrink because they emit radiation. Combining general relativity with quantum mechanics, he also declared that black holes aren't completely black and they do not live forever. As he put it: "Black holes ain't so black."

As his career was in the ascendancy, his body was weakening. He lost his ability to walk and feed himself, and his speech became so slurred that he began to use speech synthesizer, which gave him a distinctive new "trademark" voice.

Another of Hawking's trademarks was his popular approach, which invited the masses to understand scientific concepts. His introduction to cosmology, *A Brief History of Time*, has sold more than 10 million copies.

He became a pop culture icon, appearing on *Star Trek*, *The Simpsons*, *The Big Bang Theory*, and a Pink Floyd album, and was played by Eddie Redmayne in the biographical film, *The Theory of Everything*.

In recent years he predicted that artificial intelligence could spell the end of the human race. He also said that, due to climate change, humans will need to colonize another planet within one hundred years to ensure our survival.

"I think Stephen was a very normal young man," his mother Isobel said once. Mothers usually know best—but in this case, we would be justified in disagreeing with her.

BEST KNOWN TODAY FOR:
Big bangs, black holes, and much more.

Vint Cerf

(1943–)

In 1973, two decades before most of us
had even heard of the internet, Vint Cerf
invented something called TCP/IP. This
may not sound that exciting, but it is
something you probably use every day.

Born in New Haven, Connecticut, Cerf went to Van Nuys High School in California alongside future cyber geeks Jon Postel and Steve Crocker. After leaving school he earned a Bachelor's degree in mathematics from Stanford University.

Then he took an IBM job as systems engineer and focused on computer science. There, he earned a reputation as a hard worker, frequently putting 18-hour days. After two years at IBM he left to obtain an M.S. and PhD degree from UCLA, where he also worked as an assistant professor. It here that he met Bob Kahn.

It was a fortuitous meeting. Cerf and Kahn joined forces to co-design the TCP/IP (Transmission Control Protocol and Internet Protocol, respectively). This is the set of communications protocols that allows data to flow from computer to computer across the internet. Its invention paved the way for wifi, Ethernet, LANs, email, FTP, and 3G/4G. Its invention was only half the job: Cerf then had to cajole and convince user communities to adopt it.

As a result of his success, Cerf (with Kahn) became known as the "father of the internet." Yet he remembers that the foundation of the code came by "literally sketching the idea out on the back of an envelope" in a San Francisco hotel when he and Kahn were attending a computer conference.

He has since co-founded the Internet Society and served as president of ICANN, the organization that runs the domain naming system. In 2004, he was the recipient of the ACM Alan M. Turing award, which is often referred to as the "Nobel Prize of computer science." In 2005 he was awarded the Presidential Medal of Freedom.

Despite these honors, Cerf is remarkably humble. Although he gets mobbed for autographs and selfies at computer conferences, he shrugs off his importance in the history of the net. "To say I'm personally responsible for what the internet has become is not a fair characterization," he said. "A number of people, in the thousands by now, have had some kind of impact on it."

BEST KNOWN TODAY FOR:
The protocols that effectively define the modern internet.

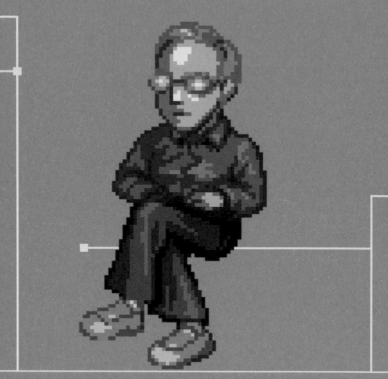

Ward Christensen
(1945–)

In January 1978, a huge snow blizzard hit Chicago. This was a major weather event, compared by one member of the national guard to a nuclear attack. Ward Christensen and his fellow computer club member Randy Suess spent the time holed up inside, developing the first bulletin board system.

He'd always loved technology. At high school, he had built a computer of sorts—a "10-stage binary counter with phone-dial input." He then, in his own words, "p**sed away" three semesters at university, unsure of what he wanted to do with his life. However, hooking up with IBM put him on the road to his great creation.

As the blizzard hit, Christensen and Suess were brainstorming a computerized message system, inspired by library bulletin boards. While Suess assembled the hardware —a basic computer with 40k of memory and a 300-baud modem— Christensen wrote the code and served as the system operator.

The project took exactly a month from conception to completion. At first it was a rudimentary affair: only one user could dial in at a time, and data transferred at about five words per second. They named it CBBS, provoking a nerdy debate on what the initials stood for. Early theories were that it was Christensen's Bulletin Board System, or Chicago Bulletin Board System. However, Ward explained that it meant Computerized Bulletin Board System, and the C was later dropped, giving us "BBS."

Most of the planet had not even heard of the internet, yet Christensen had got the ball rolling on what would become one of its most powerful and profitable dimensions.

He described his work as "a habit— like a drug," and also a "withdrawal" from society. He says this suited him because he was a "loner," who was never interested in partying. He even hated driving, because, as a "very law-abiding" citizen he found it "unbearable" to watch others break the law.

It is pleasantly ironic that his withdrawal from the world ultimately led to everyone becoming more connected than ever. For with his creation of CBBS, arguably the world's first open online social network, he had paved the way for internet forums, and then Friends Reunited, MySpace, Facebook, and Twitter. Christensen was a visionary who saw the possibilities of the Web even as it was just coming into existence.

BEST KNOWN TODAY FOR:
Taking the first steps toward internet forums and social networks with his invention of online bulletin boards.

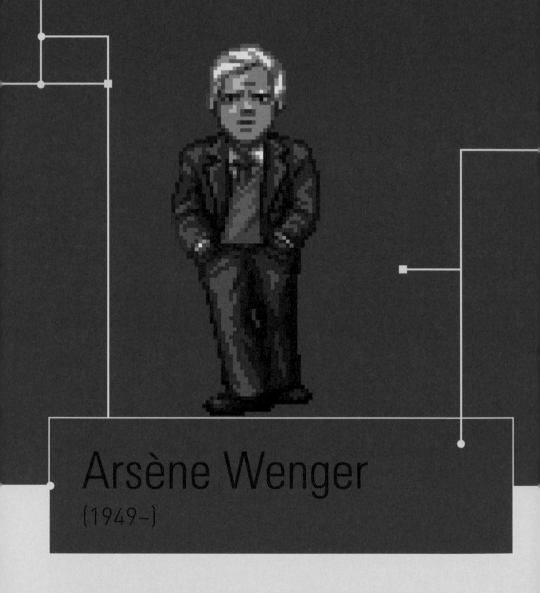

Arsène Wenger
(1949–)

"Island mentalities are historically mistrustful of foreign influences," said football manager Arsène Wenger. The parochial suspicion when the little-known French coach arrived in English football neatly demonstrates his point.

Following a spell coaching in Japan, Wenger arrived in England in 1996. Football (or soccer) was still very much a provincial sport dominated by alpha males. Among these barrel-chested boors, the lithe, bespectacled, quietly spoken Wenger stood out like a sore thumb. He spoke six different languages—French, German, English, Spanish, Italian, and some Japanese. This was quite a contrast to English players and managers, many of whom struggled to grunt articulately in their native tongue.

With piercing intelligence and scientific attention to detail, the new boss of Arsenal set about transforming the English game. On the training ground, he replaced stamina slogs with closely measured short-burst sessions. In the canteen, he replaced stodgy fare with healthy food and sporting supplements. He even chose the cutlery and chairs. On the team bus, he modified the temperature to keep the players' muscles supple. Drawing on his master's degree in Economics from his hometown university in Strasbourg, he also guided Arsenal through the prolonged period of austerity required by the building of a new stadium.

The English game had never seen anything like it, and success on the pitch for Arsenal soon followed. The team played graceful and pacy football. They won league and cup doubles in 1998 and 2002, and went an entire season unbeaten in the league in 2004. In 2017 he became the most successful manager in FA Cup history when he won the famous trophy for the seventh time.

Wenger's methods were adopted by clubs throughout the country, yet some of the game's dinosaurs continued to eye him with suspicion. He was accused of being "aloof" for declining to take part in the traditional post-match ritual of wine with the opposing manager.

A quiet, gentle, and introverted man—albeit with a witty turn of phrase—he had success coaching players who had more reserved natures, including the shy-but-sensational Dennis Bergkamp and Mesut Özil. He also became an icon for the football lovers who had always felt a bit uncomfortable amid the laddish "banter" of the beautiful game. Whichever team they support, geeky football fans will always have a soft spot for Arsène Wenger.

BEST KNOWN TODAY FOR:
Transforming English football and going an entire league campaign unbeaten with Arsenal FC.

Steve Wozniak
(1950–)

As a child, Steve Wozniak told his engineer father that one day he would own a computer of his own. "Well, Steve," replied his dad, "they cost as much as a house." Wozniak was surprised but unbowed: he spent his high-school years designing dozens of computers on paper.

The likeable, boyish "Woz" is perhaps the greatest engineer and inventor of the modern era. As well as the Apple computer itself, he invented the floppy disc, external memory, and color on computer screens. A pioneer of the personal-computer revolution, he has a net worth of $100 million, but for Woz his mission has always been about having fun.

At school he printed cards with fabricated classroom changes, plunging the entire school into chaos. He made a hoax bomb that caused the evacuation of the school, and was expelled from university after hacking into its computer system.

He later worked at Hewlett-Packard and ran a dial-a-joke company, through which he met his wife. In 1976, he was introduced to Steve Jobs, who was just the character to harness his inventiveness. Jobs noted Woz's love of fun, but asked: "How does that fit into a larger cohesive vision that allows you to sell $8 billion, $10 billion of products a year?"

Their first venture together was creating devices that enabled people to make illegal free calls. Wozniak later regretted the project, but it gave the duo "a taste of what we could do with my engineering skills and Jobs's vision."

Then they holed up in Jobs's garage. Their project: to create a user-friendly alternative to the computers of IBM. Jobs would be the marketer, Woz the inventor. He designed the Apple I computer, which came in a rough-and-ready wooden box.

Then he created the more refined and pretty Apple II, which became one of the first successful personal computers for the mass market, selling in excess of six million before production ended in 1993. "When you're starting out in a brand new field and you have a first-mover advantage, everything you touch is gold," he told Forbes. By 1983, Apple had a stock value of $985 million.

In the 1980s Woz left Apple and was injured when the private plane he was piloting crashed during takeoff. After the recuperation period, he continued to innovate. He established CL 9, the firm behind the first programmable universal remote control, created the Electronic Frontier Foundation, to provide legal aid for computer hackers, and the Wheels of Zeus, developing wireless GPS technology.

"I was lucky," he reflected. "Keys to happiness came to me that would keep me happy for the whole of my life. It was just accidental."

STEVE WOZNIAK

BEST KNOWN TODAY FOR:
His part in the Apple story; a pioneer of the personal-computer age.

Andy Hildebrand
(1951–)

As a child, Andy Hildebrand played the flute. He played in an orchestra and loved to experiment with music. As an adult, he invented technology that revolutionized music. It also divided listeners.

After he launched Auto-Tune, he reflected, "I've certainly created something people love to hate."

Though his first love was music, he took an interlude when he studied electric engineering at Chicago's University of Illinois, before going on to work in the oil industry with Exxon.

Returning to the music industry in mid-1990s, he reprised his experimentation after someone at an industry fair asked, "Well, Andy, why don't you make me a box that would have me sing in tune?" Initially, he thought, "Boy, that's a lousy idea," but nine months later he remembered the idea and thought: "You know, that's pretty straightforward to do, I'll do that."

He drew on something he had learned in his oil work. Companies would detonate charges in the ground or in the water, and use sensors to analyze the reflections to spot the oil. He realized a similar technology could analyze and modify the pitch of audio files.

Previously, producers had made pitch corrections by making the singer repeat a phrase dozens, or sometimes hundreds, of times, before patching the takes together to make one piece of music that sounded in tune. They would have been paid thousands for this labor-intensive and inefficient system. Now, the whole process could be managed at the touch of a button. "I put [them] out of business overnight," he said.

The public's first introduction to Auto-Tuned music came in 1998, when Cher released her single "Believe." The song was a smash hit, but as she asked whether listeners believed in life after love, some struggled to believe in life after Auto-Tune.

Why the controversy? Because rather than deploying Auto-Tune as it was intended, the production team decided to hurl subtlety out of the window. They went instead for an effect that is part human synthesizer, part robotic voice.

However, Hildebrand, who has recently pioneered the application of Auto-Tune pitch-processing technology to the guitar market, accepts no blame for any misuse of his musical baby. "I made the car," he says. "I don't drive it down the wrong side of the road."

BEST KNOWN TODAY FOR:
Auto-Tune: giving hope to average pop singers everywhere.

Shigeru Miyamoto
(1952-)

Shigeru Miyamoto grew up in a household with no television and few toys. So he made his own fun, fashioning puppets out of wood and string. He also loved to meander around the Kyoto countryside, exploring bamboo forests, his mind full of wonder.

One day, he ventured into a cave. It proved a pivotal experience: his excitement as he shone his torch around the passageways that led to various chambers stuck in his mind. It would later resurface to great effect.

During his school and college years, Miyamoto's creativity flourished. He loved to paint and he played musical instruments. After he graduated from college in the 1977, he asked his father to contact his friend, Hiroshi Yamauchi, who ran Nintendo, to see if he could wangle him some work.

Yamauchi was impressed with Miyamoto's ideas, which included a range of clothes hangers in the shape of animals. He got a job and within a few years was asked if he wanted create a new arcade game. He chose to accept the mission and told his friends, "You probably won't hear from me for about two or three months."

Miyamoto had always wondered why video games had no plot. He wanted to create something more imaginative than yet another shooting- or sports-based game. He found that his best ideas came during long soaks in the Nintendo office's bathtub. The product of all these baths was a game that featured an ape called Donkey Kong who had stolen the girlfriend of a mustachioed man called Mario. It proved the huge hit Nintendo had hoped for. Then his next creation, the *Mario* series, surpassed it.

He went on to launch a string of other successful games, including the *Legend of Zelda*, where his childhood cave experience came to the fore, and *Star Fox*, *F-Zero*, and *Pikmin*. His approach is to create "unique and unprecedented" games, following the principle, "As long as I can enjoy something, other people can enjoy it, too."

Miyamoto is often compared to Walt Disney and no wonder: his Mario is one of the few characters to approach the recognition level of Mickey Mouse. In 2008, readers of *Time* magazine voted him the most influential person of the year.

Miyamoto remains one of the industry's most charming characters. Despite being a god to the gaming world, he is an unassuming man, cycling to work into his fifties, until Nintendo asked their precious asset to consider a safer form of transport.

Asked which of his games' characters he is most like, he named Lakitu of the *Super Mario* series. "He seems to be very free, floating in the air, going anywhere. And that's me, that is."

BEST KNOWN TODAY FOR: Mario.

Tim Berners-Lee

(1955–)

The signs were there from the start. The son of parents who worked on the first general purpose commercial computer, keen boyhood train-spotter Tim Berners-Lee learned about electronics while fiddling with his model railway. He then made his first computer using a soldering iron and an old television set.

After studying physics at Oxford University, Berners-Lee worked at telecoms and tech companies. Then, in 1980, while working at CERN, the European Particle Physics Laboratory in Geneva, he first articulated the notion of a global system. It was based on the idea of "hypertext," which would allow people anywhere to share information.

Nine years later, he produced a paper called "Information Management: A Proposal." Here, by wedding hypertext to the internet, he invented a system for sharing and distributing information anywhere on the planet. He named it the World Wide Web—catchy initials, there.

After executing the first successful communication between a Hypertext Transfer Protocol (HTTP) client and a server, he created the first Web browser and editor. History was made when the world's first website was launched on August 6, 1991.

Its address was http://info.cern.ch—not such a catchy name, but the website concept still kind of caught on. As of September 2014 there are more than one billion websites on the World Wide Web.

Despite his epochal invention, and the knighthood that resulted from it, Berners-Lee remains a surprisingly humble man. At the outset, he refused to cash in on his invention, reasoning that: "You can't propose that something be a universal space and at the same time keep control of it."

His own website insists that no, he did not invent the internet. "The Web is an application that runs on the internet like a fridge uses the power grid," he clarifies.

He now strives to keep his vision of the Web as a "creative tool, an expressive tool" alive. He campaigns for keeping the Web open, and ensuring no single company dominates it. The World Wide Web Foundation, which he founded in 2009, bids to harness the Web's social and democratic power and to promote web access as a human right.

So what of the future? "The Web as I envisaged it, we have not seen it yet," he says. "The future is still so much bigger than the past."

BEST KNOWN TODAY FOR:
Inventing the World Wide Web.

TIM BERNERS-LEE

Bill Gates
(1955–)

Bill Gates created the world's largest company, revolutionized computers, and became the world's richest man, donating billions to charitable causes. Not bad for a college dropout.

He was born on October 28, 1955 and grew up in Seattle. As a kid he was a bookworm and a reclusive character, prompting concern from his parents who could never have imagined the feisty businessman he would become.

He began computing as a 13 year old at Seattle's Lakeside school, where he met fellow computer geek Paul Allen. By the age of 17, he had sold his first program: a timetabling system for the school, earning him $4,200. Before he graduated from Lakeside, he scored 1590 out of 1600 on the college SAT test. Then, while at Harvard, he and Allen wrote the first computer language program written for a personal computer. They established Microsoft in 1975, so-called because it provided microcomputer software.

Sensing there was money to be made in this programming racket, Gates dropped out of school and installed himself as the head of Microsoft, personally reviewing every line of code the company shipped, often rewriting when he felt it necessary. Going against the grain of the time, he saw free distribution of software as theft, and warned computer hobbyists that such behavior would "prevent good software from being written."

The big breakthrough came in 1980 when IBM signed up Gates and Allen to provide the operating system for the new personal computer. It became known as MS-DOS. Gates licenced the operating system to other manufacturers, creating a tsunami of IBM-compatible personal computers that depended on Microsoft's operating system.

Then he developed Windows, a system that used a mouse to drive a graphic interface, displaying text and images on a user-friendly screen, rather than deploying the grittier text-and-keyboard-command driven MS-DOS system.

The Microsoft company floated in 1986, raising $61 million. Gates's technological innovation, focused strategy, and aggressive business tactics had served them well. At this point, Allen stepped away, but Gates remained.

In 1987, he became a billionaire when the stock hit $90.75 a share. His worth passed the $100 billion mark in 1999. He stood down as chief executive of Microsoft in 2000, to focus on his passion—software development. Then, in 2014, he stepped down as Microsoft's chairman to focus on charitable work with the Bill and Melinda Gates Foundation, where he supports causes in the areas of global health and education.

BEST KNOWN TODAY FOR:
Guiding the modern technology revolution with Microsoft, and his work as a philanthropist.

BILL GATES

David Braben
(1964–)

David Braben is one of the most influential computer game programmers of all time. Known as the "Godfather of Gaming," in the middle of the 1980s he made the computer games industry an offer they couldn't refuse.

From a back-alley office behind a British Gas showroom in Cambridge, he changed the world.

Braben studied Natural Sciences at Jesus College, Cambridge, specializing in Electrical Science. In September 1984, he and fellow twenty-something geek Ian Bell created the first 3D home-computer game, knocking it together on a basic BBC Micro.

The game was called *Elite* but looked at through twenty-first-century eyes, it seems anything but that. This procedurally generated space-exploration venture had basic graphics, weak colors, and poorly defined features. It only needed 22k of memory—roughly the same size as a typical 21st-century email.

Yet back in the 1980s, this seminal game blew the minds of those who played it. Players would become obsessively immersed in the game, in which they would navigate a Cobra Mk III spaceship through eight galaxies, fight off space pirates, and earn money to upgrade their craft. Along the way they passed through various levels, from Harmless to Dangerous, and, eventually *Elite*. Real-life certificates would be sent to those proud gamers who made it all the way to the top.

Never had a computer game felt so real. Until its launch, the sector boasted only a handful of games. They were two-dimensional and limited in feature and appeal. *Elite* was a genre maker—the yardstick by which subsequent space-trading games would be measured.

Braben's stock in the computer games world is therefore immense. The colossally successful *Grand Theft Auto* is considered the offspring of *Elite*. Geeks sometimes hang around outside Braben's home, hoping to ask him questions. Since his 1980s heyday, he has created new games, including *Rollercoaster Tycoon*. In 2014, he produced a new game—*Elite: Dangerous*, a definitive multiplayer space epic.

Thirty years on from the launch of his first game, Braben had done it again. In the same year, he was appointed Officer of the Order of the British Empire for services to the UK computer and video games industry.

Braben is also actively trying to encourage the next generation of programmers: he is the co-founder of Raspberry Pi Foundation, a charity that has developed a series of low-cost computers to bring computer science into schools and developing countries. Bring on the next elite generation!

BEST KNOWN TODAY FOR:
Trailblazing games like *Elite*, which, with its open world, changed the face of gaming forever.

Rivers Cuomo

(1970–)

Weezer's eponymous debut album was described accurately by one critic as: "absolute geek-rock, out and proud." Indeed, it was this band and this very album that propelled geek-rock music into the mainstream. The visionary behind it was the band's lead vocalist, guitarist, and songwriter: Rivers Cuomo.

His life was a little out of step from the start. Born in 1970, his left leg was 1.7 inches (4.4 cm) shorter than his right. He grew up in an ashram in Connecticut called Yogaville. At age 11, he declared himself a metalhead and began to play Kiss covers.

He moved through a few bands before forming Weezer in 1992. Their work became a benchmark for smart, shy boys who had trouble talking to girls and dressed gawkily. Cuomo said his lyrics, which he conjured from the steadfast journal he kept, were simply a reflection of his own anguished emotional state.

Geek-rock was alive: Songs would now reference comic books, science fiction, and video games. The lyrics would be rich on irony, wordplay, humor, and self-deprecation. Casting aside the aspirational, boasting tone of other genres, geek-rock bands loved nothing more than to celebrate being a "loser."

After the success of a debut rock album, some artists dive headfirst into a life of decadent hedonism. Not Cuomo—he responded by enrolling at Harvard University to become a composer of classical music. "The only time I could write songs was when my frozen dinner was in the microwave," he said. "The rest of the time I was doing homework."

Although the 1990s were notable for producing a string of overconfident bands, made up of boorish braggarts, Cuomo was always classier and gentler. "I thought my songs were really simplistic and silly, and I wanted to write complex, intense, beautiful music," he said. Rather than becoming a rock cliché, he grew a beard, kept a diary, took a vow of celibacy, and fasted. He took up meditation and volunteered at a food bank serving HIV patients.

As the godfather of a genre, Cuomo has influenced countless bands including the Ben Folds Five, Fountains of Wayne, and Wheatus. He remains adorably nerdy to this day: When he signed up for the dating app Tinder, in 2016, his profile modestly read, "Not looking to hook up, just trying to have new experiences and get some ideas for songs."

BEST KNOWN TODAY FOR:
A legacy of geek-rock hits, including "Tired of Sex," "Undone—The Sweater Song," and indie-club favorite "Buddy Holly."

RIVERS CUOMO

Larry Page

(1973–)

In recent years, Larry Page has invested in a quest to extend human life five-fold. If ever there was a life that deserved to be extended it is that of Page himself: during the first four decades of his own existence, he changed the face of the planet.

Born in East Lansing, Michigan, in 1973, Page grew up surrounded by digital inspiration. His parents both taught computer programming at Michigan State University, on top of which his father was a genius in the fields of computer science and AI. Their home was rammed with computers and stacks of technology magazines.

At Stanford University, where he was studying computer engineering, he met Sergey Brin. They worked on an idea to rank webpages by their inbound links, instead of how many times they contained a queried term. This method would make it harder for search-engine cheaters who could trick their pages into the search results for unrelated topics merely by embedding appropriate keywords.

To make their breakthrough, they first had to download the entire internet—which caused anger when it gobbled up nearly half of Stanford's bandwidth. Then, as with so many online ventures of the time, it initially burned money at a terrifying rate. They named their concept BackRub, but later renamed it Google after the mathematical term "googol," which refers to the number one followed by 100 zeros.

After raising $1 million from family, friends, and other investors, Page and Brin launched the company in 1998. They tried to monetize it with pay-per-click advertising. It worked: the pair went from PhD students to billionaires in just five years.

Page credits the alternative teaching method of the Montessori school he attended with encouraging him to "not follow rules and orders" but to "question what's going on in the world." His business philosophies include a rejection of delegation and bureaucracy: he has been described by a colleague as "curious" and "idealistic." Under his "10X" approach, Google staff are challenged to create products and services that are at least 10 times better than those of its competitors.

But perhaps his most consequential philosophy struck him at the age of 12 when he read a biography about Nikola Tesla, who died in debt and obscurity. Page cried as he finished the book, and he concluded that inventing things "wasn't any good—you really had to get them out into the world and have people use them to have any effect."

He has managed just that. If you want to learn more—Google him.

BEST KNOWN TODAY FOR:
Google.

LARRY PAGE

Limor Fried

(1979–)

An Ada Lovelace for the twenty-first century, engineer Limor Fried is best known by her online moniker "Ladyada," in tribute to Lovelace, the world's first computer programmer. It is a bold comparison to make but one that is not without justification.

Fried has become a poster child for so many movements and trends: female entrepreneurship, women in male-dominated industries like engineering and manufacturing, and the open-source and hacker community.

Open-source hardware is hardware for which the design is made publicly available so that anyone can make, modify, distribute, and use it. The possibility of this has, to an extent, democratized the industry and has clipped the wings of the giants.

It is in this culture that Fried has most made her mark. As a child, she built, altered, and hacked electronic devices, sometimes creating her own exclusive gadgets. Then she studied Electrical Engineering and Computer Science at the Massachusetts Institute of Technology. While at MIT she founded Adafruit, an open-source hardware company.

It began when she published plans for an open-source MP3 player on her MIT webpage. She then added other electronic projects and started selling kits. By day, she did her classwork, and she worked on her new business. "I added a Paypal button, and that's how Adafruit got started," she said.

The company's other aim is to teach the world engineering, Fried sees Adafruit as "an educational company that just happens to have a gift shop at the end." Many hobbyists dream of turning their passion into a money-making venture. Limor had done it.

With her bright pink hairdo, Fried has become an iconic and recognizable figure. She was the first female engineer on the cover of *WIRED* magazine and in 2012 was awarded *Entrepreneur* magazine's Entrepreneur of the year. She was named a White House Champion of Change in 2016. Her company, Adafruit, is a 100 percent woman-owned company.

She says she wants to inspire others, to create a new generation of Ada Lovelaces. She hopes "some kid" will see her work and "start the journey to becoming an engineer and entrepreneur." For this, she realizes, she has to make the sector seem fun.

"I want to show people that engineering isn't something cold and calculated," she told the *New Yorker*. "Thinking like an engineer is a beautiful and fascinating way to see the world, too."

BEST KNOWN TODAY FOR:
Being a champion of the open-source movement, as well as a feminist icon in a field in which women and girls are massively under-represented.

Mark Zuckerberg

(1984–)

Talk about things escalating quickly. When Mark Zuckerberg developed his first Facebook prototype, he only wanted to compare students' "hotness" on his university campus. Within a little over a decade, the social network was connecting more than 1.8 billion monthly users, and Zuckerberg had a net worth of $74.2 billion.

He was born on May 14, 1984, in White Plains, New York. He loved computers as a kid—at the age of 12, he used Atari BASIC to create a messaging program he named "Zucknet," which his father used at his dental practice. When his parents hired a private computer tutor for their son, the poor teacher complained that he couldn't keep up with his young pupil.

At Harvard, Zuckerberg studied computer science and psychology. In his spare time, he co-founded a college-based social-networking website out of his college dorm room. He called it The Facebook. By 2005, it was renamed simply Facebook and had one million users.

The following year, Yahoo tried to buy the company for $1bn. Thanks to what a colleague describes as "an almost absurd degree of self-confidence," Zuckerberg rejected the offer. Another year passed and Microsoft offered $15bn—he turned that down too.

In 2012, Facebook had its initial public offering, which raised $16 billion, making it the biggest internet IPO in history. The website's reach became almost terrifying: on a single day in 2015, one in seven people on Earth used Facebook. That's one billion people.

This mammoth success story has changed our attitude to privacy: never before did we so openly share our personal details or offer up such vast data for free. It has also become a political tool: analysts say elections are often won by the side that used Facebook best. It has even been credited with boosting revolutions and civil uprisings in Europe and the Middle East.

For several years, speculation has soared that Zuckerberg will run for the White House. Though he denies the rumors, the evidence mounted quickly. He has hired a pollster and campaign managers, including an adviser to former President Barack Obama. He embarked on a year-long "listening tour," traveling to all 50 US states with a photographer who worked for both the Bush and Obama presidential campaigns.

The recent Cambridge Analytica scandal, in which Facebook was discovered to have been harvesting (and selling) the personal data of millions of its users, will likely have derailed this ambition. At least, that is, for the time being.

MARK ZUCKERBERG

BEST KNOWN TODAY FOR:
Facebook: the social network that has become perhaps the most popular and controversial plank of the internet age.

Chas Newkey-Burden is a journalist and the author of a number of books, includinging *Great Email Disasters*, a series of official titles for Arsenal FC, and biographies of Amy Winehouse, Adele, and Taylor Swift.